ACTA UNIVERSITATIS UPSALIE
Uppsala Studies in Economic Histor
Distributor: Uppsala University Libr

Carl Jeding

Co-ordination, Co-operation, and Competition.
The Creation of Common Institutions for Telecommunications.

Akademisk avhandling som för avläggande av filosofie doktorsexamen i ekonomisk historia vid Uppsala universitet kommer att offentligt försvaras i universitetshuset, sal X, onsdagen den 30 maj 2001, kl. 10.15.

ABSTRACT

Jeding, C. 2001: Co-ordination, Co-operation, and Competition. The Creation of Common Institutions for Telecommunications. Acta Universitatis Upsaliensis. *Uppsala Studies in Economic History* 55. 229 pp. Uppsala. ISBN 91-554-5008-3.

The telecommunications sector is an example of an industry that requires extensive co-ordination of e.g. technological, economic, and administrative factors in order to function as a unified system. The different actors in the sector normally have diverging interests concerning the choice of common rules. This has created a demand for institutional and organizational structures that can help the actors reach common solutions.

This study shows that co-ordination through an international organization, CCIF, created a stable institutional environment for the participating actors. It also defined which type of behaviour was profitable for states and organizations that wanted to influence the CCIF's decisions. The most central actors of the CCIF were those that participated heavily in the organization's Plenary Assemblies and expert study groups.

In this thesis it is shown that co-operation between the Scandinavian countries through informal agreements expanded and intensified both in scale and scope between 1900-1960. The individuals taking part in those agreements showed a very stable pattern over time. This indicates the evolution of a social network between the actors and organizations.

On the national level the state had the formal power to impose its will on all the other actors. The liberalisation in the 1990's of the Swedish telecommunications sector led to the creation of a large set of new rules. This study shows that the need for a new regulatory function was coupled to the political objectives for the sector defined by the state.

The results show that it is insufficient to regard the co-ordination of telecommunications as purely technical exercises. The strategic interests of actors were often based on other policy objectives than technical considerations.

Keywords: Telecommunications, regulation, international agreements, liberalisation.

Carl Jeding, Department of Economic History, Uppsala University, Box 513, SE-751 20 Uppsala, Sweden

ACTA UNIVERSITATIS UPSALIENSIS
Uppsala Studies in Economic History, 55

Carl Jeding

Co-ordination, Co-operation, and Competition

The Creation of Common Institutions for Telecommunications

Uppsala 2001

Dissertation for the Degree of Doctor of Philosophy in Economic History presented at Uppsala University in 2001

ABSTRACT

Jeding, C. 2001: Co-ordination, Co-operation, and Competition. The Creation of Common Institutions for Telecommunications. Acta Universitatis Upsaliensis. *Uppsala Studies in Economic History 55*. 229 pp. Uppsala. ISBN 91-554-5008-3.

The telecommunications sector is an example of an industry that requires extensive co-ordination of e.g. technological, economic, and administrative factors in order to function as a unified system. The different actors in the sector normally have diverging interests concerning the choice of common rules. This has created a demand for institutional and organizational structures that can help the actors reach common solutions.

This study shows that co-ordination through an international organization, CCIF, created a stable institutional environment for the participating actors. It also defined which type of behaviour was profitable for states and organizations that wanted to influence the CCIF's decisions. The most central actors of the CCIF were those that participated heavily in the organization's Plenary Assemblies and expert study groups.

In this thesis it is shown that co-operation between the Scandinavian countries through informal agreements expanded and intensified both in scale and scope between 1900-1960. The individuals taking part in those agreements showed a very stable pattern over time. This indicates the evolution of a social network between the actors and organizations.

On the national level the state had the formal power to impose its will on all the other actors. The liberalisation in the 1990's of the Swedish telecommunications sector led to the creation of a large set of new rules. This study shows that the need for a new regulatory function was coupled to the political objectives for the sector defined by the state.

The results show that it is insufficient to regard the co-ordination of telecommunications as purely technical exercises. The strategic interests of actors were often based on other policy objectives than technical considerations.

Keywords: Telecommunications, regulation, international agreements, liberalisation.

Carl Jeding, Department of Economic History, Uppsala University, Box 513, SE-751 20 Uppsala, Sweden

© Carl Jeding 2001
ISSN 0346-6493
ISBN 91-554-5008-3
Printed in Sweden by Elanders Gotab, Stockholm 2001
Distributor: Uppsala University Library, Box 510, SE-751 20 Uppsala, Sweden

To the loving memory of
Gunnar Jeding and Gösta Höglund

Contents

Acknowledgements

Working with this thesis I have received a tremendous amount of help and good advice from a large number of people. Although I can not fully repay their services, at least I can mention some of them and in this way express my gratitude.

Assistant Professor Jan Ottosson has been my thesis advisor since the beginning, and his efforts have gone far beyond the call of duty. At odd weekdays, hours, and in odd places he has always been ready to discuss my ideas, read my papers, and clarify my thoughts on short notice. I am lucky to have had him as my advisor.

Professor Lars Magnusson has on a number of occasions read large piles of my manuscripts and always offered in-depth, constructive suggestions that have seriously improved the finished product. He has also managed the research project *"Transports and Communications in Perspective"*, in which I have worked during this whole period, thus creating stable working conditions that are rare to most PhD students. Within that project I have also had the good fortune to work together with Magnus Carlsson, Rikard Skårfors, and Eva Liljegren who have been my colleagues and friends. Other participants in that project have been Professor Olle Krantz, Assistant Professor Lena Andersson-Skog, and Thomas Petterson, all three at the Department of Economic History at Umeå University.

People at the Department of Economic History, Uppsala University, have generously shared their time and thoughts. I thank them all, but would like to specially mention Anders Sjölander, Professor Kersti Ullenhag, Professor Maths Isacson, Mikael Lönnborg-Andersson, Hilda Hellgren, Lynn Karlsson, and Birgitta Ferm.

One important period for me in this work was spent at St Antony's College, Oxford. My sincere thanks to James Foreman-Peck, Mark Hickford, and the participants at the Graduate Workshop in Social and Economic History at Nuffield College. Special thanks also to Carol Brown who is a dear friend and who has persisted in trying to teach me proper English.

Over the last four years I have also had the opportunity to try my ideas at a series of workshops on the regulation of infrastructure, held jointly between Uppsala University and Northern Jiaotong University, Beijing. For their comments and suggestions I would specially like to thank Professor

Rong Chaohe, Professor Zhao Jian, Huang Haibo, Guo Wenling, Daniel Hallencreutz, Jenny Andersson, Mattias Burell, and Sofia Murhem.

Since 1998 I have divided my time between doing research and working at the Swedish National Post and Telecom Agency, PTS. This would not have been possible without the understanding and interest shown by Nils Gunnar Billinger and Laila Linnergren-Fleck. My thanks go to them, too.

This work could not have been made without financial support. The Swedish Transports and Communications Research Board (KFB) has generously financed my research from day one. The British Council helped me, through the Prince Bertil Scholarship for 1996, cope with the extra costs associated with studying one year in Oxford. The Swedish Foundation for International Cooperation in Research and Higher Education (STINT) financed the research co-operation between Uppsala University and Northern Jiaotong University. Thanks to them all.

Thanks to the Scandinavian Economic History Review and to Edward Elgar for their kind permission to reprint Jeding, Ottosson & Magnusson (1999), and Jeding (2001) respectively.

Finally, my warmest thanks to all friends and family who have remained encouraging for over six years. Most of all to my wife, Kerstin, who has shared all the ups and downs and who is always my best support.

Stockholm, March 2001

Carl Jeding

Contents

Introduction

A classic problem in the social sciences is how to achieve unity in a system where a large number of autonomous actors have diverging interests, but where it is in the collective interest to reach a common solution.

Telecommunications, being a network industry, is a very clear example of this need for co-ordination. Technologically all the parts of a telecommunications system have to be compatible in order for the system to work as a unified whole. But not only technological factors have to be compatible. Also administrative practices and economic agreements are needed for the interconnection of the various parts of the system that has been described as "the world's largest machine"[1].

This thesis studies three different cases of how actors have solved this problem of co-operation in telecommunications: through the creation of an organization entrusted with finding common rules, through voluntary agreements, and through regulation imposed by the state.

The three different examples of mechanisms for establishing common rules for the telecommunications system that are presented here are all concerned with the basic problem of private and public interests. Key questions regard the construction of organizations and institutional arrangements with the purpose of achieving co-ordination within the system, and overcoming the differences in objectives between its actors. How were such institutional arrangements made? Which were the driving forces, and who were the driving actors behind them? How did they work? And how was the development of institutions related to the actions of the actors?

The aim of this thesis is to analyse how institutional arrangements have been designed to cope with the problems that arise when the interests of concerned actors have been diverging.

One solution to the co-ordination problem is to create an independent actor with the specific task to create common rules, with the underlying idea that this actor will put the collective interest of reaching a common solution before any partisan interest of the concerned players. On an international level, the problem of co-ordination of national telephone systems was an acute problem in the early decades of the 20th century. European states had built up their national telephone networks independently of each other. This

---

[1] Karlsson & Sturesson (1995).

had given rise to divergences between them on technical aspects as well as administrative practices. Furthermore economic agreements between the European states were needed in order for them to interconnect their networks and allow the technological advances of long-distance telephony to be realised on the European continent.

With the expressed aim of facilitating such agreements between countries in the interest of creating a working, coherent European telephone network, a group of European states in 1923 formed an international organization known as CCIF[2]. Two of the articles in this thesis study the creation of common international rules through the CCIF, and how member states acted strategically to influence these common rules with their own preferences.[3]

A second solution to the co-ordination problem is to establish a forum where the different actors can discuss common problems and reach voluntary agreements. Jeding, Ottosson, & Magnusson (1999) study the case of Scandinavian co-operation in a number of network industries during the period 1900-1960. One of the findings in that study is that once a forum for discussions was established to deal with issues where the states had common interests, the very act of co-operation seemed to reinforce itself. As the telecommunications Administrations of the Scandinavian countries had established a channel for co-operation, the issues they worked together on gradually expanded from rather moderate beginnings into a very far-reaching and intimate collaboration. And through this deepened contact, the interests of the states gradually came to overlap more and more.

A third example of reaching common solutions to the co-operation problem is found on the national level. There one hierarchically superior player, i.e. the state, can enforce rules and regulations that are binding for the other actors. This way of problem solving is typically not available on the international level, but on the national arena the state has the power and function to establish institutions that lead to coherence of some kind in the system. Jeding (2001) studies the case of the liberalisation of the Swedish telecommunications market in the 1990's. In that process of regulatory reform, the Swedish telecommunications system went from being essentially an unregulated monopoly into a regulated open market. This is an illustration of the fact that the need for co-ordination increases as the number of players increases.

The object of study is the changes in the amount of regulatory activity, and in the forms of regulation that followed from the liberalisation of the sector. This is distinct from many other studies of the liberalisation of

---

[2] The abbreviation stands for *Comité Consultatif International des communications téléphoniques à grande distance en Europe*. The organization still exists and is today known as ITU-T.

[3] See Jeding (1998); Jeding (2000).

Swedish telecommunications, which focus rather on the political or corporate strategies leading to regulatory reform.[4]

Under the old system the national state-owned operator, Televerket, had a *de facto* monopoly on telecommunications in Sweden. The state could then use its direct control over Televerket as a means of directing the Swedish telecommunications sector. With the opening of the Swedish market to other operators, Televerket could no longer play that role but had to be able to act as an independent operator. To ensure neutral rules and create conditions for working competition in the market, the Swedish state then introduced a regulatory package which included a changed status of Televerket into a public company, Telia; a new Telecommunications Act; and a new independent regulatory authority, PTS. In this way, the liberalisation of the Swedish telecommunications market did not mean the abandoning of rules for the sector, but instead the creation of a whole new set of institutions.[5]

Telecommunications as an object of study

In this thesis these issues are studied within the telecommunications industry. The reasons for this choice of sector are suggested above. The telecommunications sector is a network industry, and as such its operation is based on a network of nodes. The concept of 'technical linkage' indicates the importance for the function of the network as a whole that all its parts can work together.[6] With such a concept in mind, the telephone system is even more tightly coupled than many other network industries such as for instance airline traffic or postal services. In other words, in the telecommunications sector there is and has been since its inception a strong demand for coordination between its different parts.

A second reason for the chosen object of study is that for the largest part of its history, and in most countries of the world, telecommunications have been state owned and/or state controlled. This makes the link very obvious between politics on the one hand and technological and economic development on the other.

The background for the heavy state involvement in the sector is to be found in a number of different motives. Infrastructural systems in general have often been considered key national resources where the state has found reason to intervene, with the motivation that they are important for the national economy as a whole. Over the 20th century, the relative importance of telecommunications to that of other infrastructural systems increased, so that

---

[4] E. g. Ioannidis (1998); Karlsson (1998); Berg (1999).

[5] Jeding (2001).

[6] The concept of technical linkage is used for instance by Perrow (1984); Kaijser (1994; 1999).

what was true in general for such systems was particularly true for telecommunications. Another aspect of this argument of the national importance of telecommunications is the necessity for the state of having access to a stable system of communications in times of war and other types of crisis.[7] The sector has also held a symbolic value for states in the sense that a well-developed telecommunications system has been a sign of being an economically and technologically advanced nation.[8]

Although the regulation of telecommunication has become a widely researched issue during the last decade or so, these research efforts have been mainly directed at the theoretical, and the present state of the sector. The history of telecommunications holds large gaps where research is still missing. This gives a third reason for the choice of sector. It also provides a contrast to, for example, railways which share many of the characteristics of telecommunications; for instance being a tightly coupled system with large needs for co-ordination, and a heavy involvement by the state for essentially the same reasons etc. In the case of railways however, they have been widely researched by economic historians for a large part of the last century.

The strong involvement by the state has made visible the political aspects of telecommunications regulation. This is not to say that the institution building would have been non-political if the actors had all been private.[9] However, since the national interests in international agreements and organizations were in most cases represented by the respective telephone Administrations, this points directly to the importance of political considerations in the formation of national positions.[10] The fact that the management of the national telephone systems was also a question of national political sovereignty gives further poignancy to the basic problem of finding common solutions to issues where the actors all have different interests, but need to co-operate in order to have the benefits of a working system.

Previous research

As the relative importance of telecommunications in the world's economies increased during the last decades of the 20th century, there was a virtual explosion of research on different aspects of these systems. In the following I will try to give an outline of some of that research, which is directly pertaining to the above mentioned principal interests of this thesis. The discussion follows the three broad types of solution to the problem of co-ordination

---

[7] Headrick (1981), ch. 11; Foreman-Peck & Millward (1994).

[8] Helgesson (1995).

[9] Two interesting accounts of the highly political activities of the American telephone operator ITT are given in Sampson (1973) and Sobel (1982).

[10] Jeding (1998), pp. 105-107.

that is studied here: co-ordination through international organizations, through voluntary agreements, and through national regulation.

Co-ordination through international organizations

The setting of international standards and negotiation of international agreements in telecommunications is an aspect of international relations rarely studied by those who write on political and economic relations between states. To some extent this is understandable since those agreements in the telecommunications sphere seldom give rise to high profile international conflicts. One notable exception is the allocation of radio spectrum frequencies to countries, where each country is allocated frequencies in relation to their political, social, economic, and military needs. Another exception above all during the cold war was the issue of harmful interference of radio and broadcasting. But apart from these rare examples, international regulation of telecommunications has seemingly been a harmonious field of international co-operation.[11]

If students of international politics rarely discuss telecommunication rules, the reverse is also true. That is, those who have studied the history of international rules and regulations of the telecommunications sector almost unanimously emphasise the co-operative, non-conflictual side of those relations. Moreover, they surprisingly often emphasise the technical side of the process and deny any political and economic aspects of it.

To a large extent this might be explained by the fact that most of those earlier studies of international telecommunications agreements have been written by engineers who were part of that history themselves.[12] To some extent an explanation can also be derived from the involved participants' self-interest. The more the participating engineers can convince other actors that what they do is make complicated technological decisions and not political or economic ones, the more they can be left alone to make those decisions. This is illustrated by one engineer writing about the shared project between the British Post Office and AT&T to design and build the first Trans-Atlantic telephone cable in 1953. "[It] provides an encouraging example of the way in which international conferences can succeed if only the politicians can be kept out of them."[13]

Codding Jr (1952) also describes the ITU between 1865 and 1952 as a history of co-operation rather than conflict. The central question he poses is: How has the ITU managed to function as an international organization over such a long period, marked by international conflicts and two World Wars? He finds the answer in the non-political nature of its work, and the fact that

---

[11] Jeding (1998), p. 13.

[12] See for instance Chapuis (1976); Valensi (1929; 1956; 1965); Fossion (1938).

[13] Clarke (1958), p. 159.

its work had almost exclusively been carried out by technical telecommunications experts.

He further notes three factors that were crucial for the efficient operation of the ITU. First, the large and increasing economic interdependence of European states, and the high density of population gave rise to a high demand for communications. That in turn created a demand for co-ordination of the kind that only an international body such as the ITU could provide. Secondly, he cites the fact that the European operators in most cases were state agencies. This meant that the implementation of common rules became a matter of national pride once they were agreed upon. His third factor deals with the institutional structure of the ITU. Aspects of the international regulations that were of a diplomatic/political or more permanent nature were dealt with at rather sparsely occurring plenipotentiary conferences. These meetings dealt with the long-term, stable rules governing the telecommunications system. More detailed technical matters were dealt with in specialist committees, constituted by technical experts from the member states' telecommunications Administrations.[14]

Another study of the ITU's history by Grimm (1972) departs only slightly from this view of the organization's work as one of co-operation on purely technical matters. He sees the ITU as performing basically a job of harmonious technical co-ordination, although political issues are sometimes injected into its work such as allocation of radio spectrum to countries. These issues are however the exceptions rather than the rule, much thanks to the participants' perception of their own work as non-political. In that way they have managed to keep out of disruptive political controversies.[15]

Lee (1996) goes further in studying the political context of the ITU's history. She finds the founding principles of the ITU in accordance with the liberal climate in European international relations during the middle of the 19th century. As this political climate changed towards more protectionist and nationalistic ideas late in the century, the ITU took on a new role as a colonialist vehicle. One example of this is found in the ITU's colonial voting rule, which allowed member states to vote for territories under their rule. Another example of this is found in the shape of the international telegraph network before the Second World War. In that system the European colonial powers functioned as central hubs in relation to the more peripheral colonies, which had little communications directly between them.[16]

Schmidt & Werle (1998) take their analysis of the political and economic aspects inherent in technical standardisation further. They set out by describing standards as "…highly visible examples of the social shaping of technology. As solutions to co-ordination problems, they reflect the multiple

---

[14] Codding Jr (1952).

[15] Grimm (1972).

[16] Lee (1996), p. 60.

social relations involved in technical systems."[17] In other words, it is misleading to regard choices of technical standards as isolated technical matters.

In cases where national representatives carry out the co-ordination process, such as in the ITU and CCIF, this opens the door to political considerations that are not simply aimed at solving problems of compatibility. Instead these representatives tend to regard national standards as strategic elements in competition between nations or national systems. "This suggests that relative-gains orientations prevail among the relevant players – that is, the actors not only want to improve their position; they also want to gain more than the others."[18]

International co-ordination through voluntary agreements

Much of the actual co-ordination between national telecommunication systems has traditionally taken place through direct bilateral or multilateral agreements. Despite this, relatively little has been written about such agreements in any systematic manner. Most of the literature on international telecommunication agreements focuses on those created within organizations of different kinds.

Monographs of different telecommunication Administrations often include references to treaties and agreements with other states, but these are seldom coherently analysed as a distinct way of solving a co-ordination problem.[19] Other studies of international telecommunications mention direct agreements between states, but do not typically treat them as objects of study *per se*.[20] Bobroff (1974) is more of a catalogue of bilateral and multilateral telecommunication agreements between the United States and other countries, than any attempt to analyse their function or creation.

Keohane (1999) writes about international regimes, that is, sets of rules of international relations that are distinct from the states or organizations that operate within the rules. In accordance with ideas of new institutional economics, such regimes are seen as institutions that restrict the operational sovereignty of states, and thereby lower the transaction costs of international relations.[21]

From a transaction cost perspective however, the costs of monitoring and enforcing are typically very high. How come then that international agreements exist and are followed? Keohane suggests that networks of international professionals, which gather information, set standards and interpret rules, play a crucial role. Networks of individuals that are stable over time,

---

[17] Schmidt & Werle (1998), p. 5.

[18] Schmidt & Werle (1998), p. 267.

[19] See for instance Heimbürger (1953; 1974); Israelsen (1992); Rafto (1955); Pitt (1980).

[20] See for instance Headrick (1981; 1991); Noam (1992).

[21] Keohane (1999); see also North (1990), pp. 3-5.

reduce the scope for opportunistic behaviour of states and thereby promote co-operation. This lowers the transaction costs, thereby making it easier for states to identify common interests and solve problems of co-ordination.[22]

National regulation of telecommunications

The enormous increase in interest of social scientists in the telecommunications sector is to a large extent explained by the reform of the sector that has taken place in most of the industrialised world since the mid-1980's. This reform process, although part of a larger international movement, has taken place on the national arena. Therefore, much of the vast literature on regulation of telecommunications specifically concerns national telecommunication reforms.

Much of this research has been carried out by economists, studying economic aspects of the telecommunications market with specific regard to different factors, which separate the telecommunications industry from other sectors of the economy. De Bijl (2000) gives an overview of recent work that studies specifically the issue of competition between telecommunication operators. Such studies typically investigate effects on competition of certain market features in models of unregulated markets. The various market features studied include network interconnection, heterogeneous demand of consumers, and price discrimination on call termination. While producing valuable insights into the basic nature of competition in telecommunications, the level of abstraction in such models often give them only limited applicability to real-life policy and regulation.[23]

Laffont & Tirole (2000) give a more comprehensive overview of economic theory relating directly to telecommunications. The authors give a background to the key economic features of the telecommunications sector, but also proceed to draw policy implications of how to achieve an efficient competition. Their work gives a very valuable contribution to the discussion of which regulatory measures can be expected to give the desired outcomes from a number of different perspectives. For instance they reach conclusions on the effects on competition of regulation that ensures non-discriminatory prices, asymmetric regulation, and regulatory issues concerning two-way access (following for instance local loop unbundling and the introduction of alternative forms of access to the consumers).[24]

The strength of this kind of works however, lies mainly in their provision of insights into how regulation should be designed to achieve macroeconomic efficiency. The various other objectives of different actors in the sector are typically not treated. Neither do such studies normally discuss the

---

[22] Keohane (1999), pp. 240-243.

[23] De Bijl (2000), pp. 25-30.

[24] Laffont & Tirole (2000).

16

organizational set-up in the sector, and the effects on the institutions resulting from it.

Telecommunication being a traditionally heavily regulated part of the economy has meant that the state has had an important role in the co-ordination of the national systems. In contrast to the mutual sovereignty of actors in international co-ordination, the state has the power to impose regulation on other actors on the national level. This has led to a large body of research concerning the state's motives for regulation. The explanations of how regulations arise, develop, and decline vary greatly among the writers. A common way of grouping these different approaches is however to focus on the motives for and driving forces behind regulation. With that classification scheme, three broad categories fall out: those that focus on public interest; on private interest; and on institutions.[25]

Public interest theories
Public interest theories set out from the notion that regulations are created to achieve some publicly desired goals that, for instance the market would not reach spontaneously. From this follows that the actors who propose and enforce regulations do so out of a public interest.[26]

Within this tradition explanations of why regulations are introduced often focus on market failures of different kinds, large asymmetries of power or information, or large externalities. This, then, would go some way towards explaining why infrastructure sectors such as telecommunications, with its large externalities and, at least up until fairly recently, natural monopoly features, has been such a strictly regulated sector of national economies. Other explanations go outside the more strictly economic justifications for regulation and focus on judgements that services or products are hazardous to the health, safety, or morals of the population.[27]

There are, however, a number of fundamental problems with the public interest approach. One of these is that it is most difficult to find an agreed notion of what really constitutes the public interest. And even if such a definition can be found, the approach also stipulates that the regulators have the necessary competence and integrity to translate this public interest into actual regulations.

---

[25] For an overview of some of the main schools in regulatory theory, see Baldwin & Cave (1999), who use a somewhat different classification. Dixit (1996) uses the broader classification scheme of normative and positive models of explanation. Laffont & Tirole (2000) use roughly the same two categories of explanations, but use the terms 'public interest' and 'political economy' approaches. Magnusson & Ottosson (2000) use approximately the same categorisation scheme as here, under the headings of normative studies, positive theories, and new institutional economics.

[26] See for instance Francis (1993), pp. 1-8.

[27] Francis (1993), p. 10.

Several studies indicate the great variation of governance structures at different points in space and time as an indication that public interest in itself is not a satisfactory explanation for how telecommunications systems are governed. Due to the incompleteness of the political contracts that make up regulation, there is large variation in the implementation of rules across countries and periods of time. This suggests that there is room for self-interested actors to influence the governance of the telecom sector. [28]

If telephone regulations are deeply embedded in other economic, technological, and political issues it becomes very difficult indeed to define a public interest that is common to all actors.

Private interest theories

Setting out from a wholly different idea of how and why regulations occur, private interest theories focus on the private interests of actors to explain why regulatory systems exist and look the way they do.[29] This approach was pioneered by the public choice school in the 1960's, and still commands much attention in the field of regulatory theory.[30] In an attempt to bring political issues such as regulation in line with neo-classical microeconomic theory, writers such as George Stigler and Sam Peltzman argue that where there is a failure of competition or a monopoly, there is also a monopoly profit. If a regulator controls this monopoly profit there is an incentive for actors to influence the regulator so as to benefit from the resulting 'regulatory rent'. Thus there is a market for regulations in rather the same way as for other commodities.[31]

This idea further implies that the commodity of regulation, in effect the regulatory rent, goes to the actors who value it most. This, then, would typically mean that regulators become 'captured' by industry preferences, since industry actors typically have more to lose or gain from regulation than regulators.[32]

Further, it means that due to asymmetric allocation of power and information, compact and well-organized interests, such as producers, typically would have an advantage over more diffused interests, such as consumers. The regulatory rent from a regulation that, say, preserves a monopoly, is shared between a relatively low number of producers, whereas the costs are

---

[28] See for instance Schneider (1991), pp. 23-29; Levy & Spiller (1996).

[29] A review of the most important foundations of what is called 'the economic theory of regulation' and capture theory, or ET and CT for short, is given in Peltzman (1989).

[30] For a good introductory overview of the public choice school, see Tullock (1994). For a more modern example of economic historians' studies of interest groups and their influence over regulations, see Goldin & Libecap (1994).

[31] Stigler (1971), p. 3.

[32] Peltzman (1989), p. 175.

distributed among a large number of consumers whose individual shares of the costs, and therefore incentive to counteract, are relatively small.[33]

This rather large body of theories of course also comes with a lot of problems. First of all there are the usual problems with neo-classical micro-economics, such as the assumptions of rational, well-informed actors with stable and consistent preferences, low or no transaction costs etc.[34]

There are also other problems, caused by events rather than theoretical considerations. The wave of liberalisation of markets that has swept the Western world since somewhere around 1975 is hard to explain through this approach.[35] Another problem is the one raised by comparative studies. If the commodity of regulation is distributed through a simple market mechanism where the actors who value it most will end up controlling it, then how come the same issues with roughly the same interest formations give such drastically different outcomes in different countries?[36] For instance Anne O. Krueger's (1996) study of American sugar regulations further shows empirical results that seem to contradict what would be predicted by pure private interest theory. In that study, interest groups with apparently opposed interests were seen to co-operate. One of the conclusions drawn is that neither pure public interest, nor pure private interest can explain the development of regulatory regimes. Instead regulatory programmes seem to gather a momentum of their own, once initiated, and gradually come to distance themselves from the preferences of their initiators.[37]

Institutional theories

Institutional theorists are normally sceptical about the atomistic assumptions of many private interest explanations, and focus instead on institutional structures and arrangements as important factors in shaping regulations. Individual actors and their behaviour are seen as influenced by rules of various kinds, ranging from legal frameworks to organizational rules and more informal, social codes of conduct.[38]

From a transaction cost perspective based on contract theory, Dixit (1996) points to the incompleteness of contracts. For a number of reasons, there is a gap between the state's intentions, as expressed in policy decisions,

---

[33] Peltzman (1989), pp. 172-8.

[34] Hodgson (1988), ch. 4.

[35] Although one explanation offered has been that the regulations themselves had become an obstacle to firms, so that it was rational for them to use their influence to liberalise markets. Peltzman (1989), pp. 38-9.

[36] See for instance Immergut (1992); Vogel (1996).

[37] Krueger (1996).

[38] Baldwin & Cave (1992), pp. 27-31.

and their implementation.[39] This gap gives room for some freedom of interpretation on behalf of actors. The resulting policy acts can, and often do,
have effects that shape future room for decisions, so that economic policy
making could be described as a dynamic game where the rules are partially
made by the participants as they go along.[40] Dixit seés economic policy
making as an evolutionary game. He thus presents a synthesis that acknowl-
edges actors' freedom to act according to their interests within a given scope
defined by rules, and the fact that rules and history shape the available alternatives to actors. In a review of the regulatory history of a number of Swedish network industries, Magnusson & Ottosson (2000) support the notion of
a difference between political decisions and their implementation, and points
to the importance of applying a historical perspective in order to understand
strategic action within this field.[41]

Regulatory reform of Swedish telecommunications

Regarding the Swedish telecommunications sector a number of studies have
been made of its reform process and of the state's instruments for controlling
the sector. Karlsson (1998) gives a thorough presentation of the process
leading up to the introduction of a new liberalised telecommunications regime in Sweden. In his effort to explain the dynamics behind the Swedish
liberalisation he finds that pure private or public interest are insufficient to
explain the outcome of the 'game' of regulatory reform. Institutional factors
are also needed to explain, for instance, why the liberalisation process was
so fast in Sweden compared to many other countries.[42]

Karlsson also differs from many other social scientists trying to explain
this type of processes in his strong emphasis on technological factors. As a
result of the technological development of the telecommunications, com-
puting, and office electronics sectors, those systems gradually converged and
a number of technical connection situations arose. Each of these connection
situations produced a conflict of interest between two sociotechnical cul-
tures, and opened part of the old monopoly system for political and strategic
reconsideration by the concerned actors. Gradually and step-by-step the
whole of the telecommunications sector was in this way opened for political
reconsideration. As the main players in the regulatory game towards the end
of the period became in favour of liberalisation, this resulted in reform of the

---

[39] See Dixit (1996), ch. 2; See also Laffont & Tirole (1993), ch. 1, on the constraints that
prevent the regulator from implementing his or her preferred policy.
[40] Dixit (1996), pp. 29-31.
[41] Magnusson & Ottosson (2000), p. 198.
[42] Karlsson (1998), pp. 309-311.

whole Swedish telecommunication sector into one of the most liberalised in the world.[43]

Berg (1999) studies the liberalisation of a number of Swedish network industries between 1976 and 1994. Regarding telecommunications he sees the liberalisation as a three-dimensional development. First, the state-owned monopoly operator Televerket/Telia gained *autonomy* from direct control by the state. Secondly its tasks became more *focused*. As the liberalising measures proceeded Telia's tasks gradually became strictly limited to those of an independent telecom operator, while other, more policy related, authority tasks were transferred to other bodies. Thirdly the telecommunications sector was opened to *competition*.[44] As also suggested by Karlsson (1998), Berg finds that a key role in the liberalisation process was played by the Social Democratic party. Being in power during much of the actual period, their attitude towards liberalisation of markets was crucial. In his study Berg shows that the Social Democratic position changed step-by-step from an initial resistance towards liberalisation and market ideas to a position where many of the liberalising policy decisions were actually taken by Social Democratic governments.[45]

Ioannidis (1998) describes the same process from the perspective of the monopoly operator Televerket, and its strategic actions to gradually gain autonomy from the state. Mölleryd (1999) deals specifically with the development of mobile telephony in Sweden, but discusses to some extent the regulatory reform concerning that part of the sector. Apart from these studies, there are a number of government inquiries and audit reports, which briefly cover certain aspects of the policy development in the telecommunications sector.[46]

Helgesson (1999) is a rare example of studying the development of a governance system for the Swedish telecommunications sector in a more distant historical perspective. In his study he investigates how the Swedish telecommunications sector went from a state of competition into monopoly. His main focus is how, during the stabilisation of that monopoly in the period 1903-1930, telecommunications gradually came to be defined as a 'natural monopoly'. Skårfors (1997) gives further detail to one facet of that picture: how the Swedish state acted to create its *de facto* monopoly by purchasing private telephone networks. Andersson-Skog (1999) also provides a brief account of the development of early Swedish telecommunications policy.

---

[43] Karlsson (1998), pp. 320-323; 309.

[44] Berg (1999), ch. 7.

[45] Berg (1999), pp. 253-258.

[46] Riksrevisionsverket (1995); Riksdagens revisorer (1995/96:5); Ds 1996:29; Ds 1996:38.

The articles

This chapter will briefly go through the four articles that are included in this thesis and summarise their results. The first two of the articles deal with co-ordination of the international telephone network through the organization CCIF. Jeding (1998) studies more specifically British strategies in the CCIF. Jeding (2000) analyses which member states, organizations, and individuals were central in the CCIF's work through studying the structure of the co-operation within that body. Jeding, Ottosson & Magnusson (1999) focuses on international co-operation outside formal organizations by studying the example of Scandinavian agreements in telecommunications matters between 1900 and 1960. Jeding (2001) leaves the international level, and studies regulatory reform of the Swedish national telecommunications sector.

This introduction is followed by a section that provides a background description of the CCIF and its organizational design. Section 2 deals with the methodology of the articles, and section 3 discusses the sources used in them. In section 4 the results of the articles are presented, whereas the main conclusions of the different studies are drawn in section 5.

Background: the ITU and CCIF

As telephone technology improved during the first decades of the 20th century, telephony over long distances became technically feasible. In the United States a network for cross-continental telephony was built up, and in the beginning of the 1920's it was possible to make phone calls between all parts of the country.[47] In Europe however, this development was hindered by the fact that telephone networks were operated on a national scale. As these national networks had been built up, more often than not in the hands of the state, diverging technological and administrative styles had taken shape, making interconnection between the national systems troublesome or impossible.[48]

---

[47] Jeding (1998), p. 7; p. 40.

[48] Jeding (1998), pp. 31-33. On the issue of technological style, see Hughes (1987), pp. 68-70. On national differences in governance of telecommunications systems, see e.g. Schneider (1991); Noam (1992); Levy & Spiller (1996).

First of all the national telephone systems needed to be technologically compatible so that a call could technologically be carried between them. The national Administrations also had to agree on minimum standards of maintenance in order to ensure the reliability of the system. The quality of the service depended on the quality of the whole network, so that a call between, say, London and Berlin depended on the quality of the systems in all the transit countries in between them. An international telephone call required being prepared completely from one end to the other before any actual conversation could begin. In the days of manual switching precise standards were needed to regulate how a call should be set up, in order to minimise time 'lost' in administrating the calls. In addition to this the application of tariffs and the collection of fees had to be agreed upon in order to establish the international telephone service.[49]

To overcome these difficulties a group of European states, on a French initiative, formed an international organization called CCIF. Its purpose was to find common technological and administrative standards and practices in order to co-ordinate the national telephone networks into a single, coherent international system.

The CCIF was organized as a forum for co-operation between the respective national telephone administrations. Harmonisation of rules and standardisation of technology and practices was seen as something in the interest of each individual member state as well as in the common interest of all. The chosen organizational form was however institutionally weak. Participation in the CCIF was voluntary, and the common standards and practices that the delegates decided on were issued as recommendations to the member states. Despite this, the CCIF had a powerful impact on the interconnection of the previously almost isolated national telephone systems in Europe. During the first years of the CCIF's existence, the organization quickly gathered new member states, and a large number of new telephone connections opened across Europe.[50]

The CCIF's organization built on three bodies: the Plenary Assembly, the Committees of Rapporteurs, and the General Secretary. The Plenary Assembly was the executive body. It met once a year (after 1932 once every two years) and consisted of delegations from those national Administrations or companies that were a member of the CCIF. Only one delegation per country was accepted, so in the cases where several operators took part from one country, these had to be co-ordinated under one head of delegation. The delegations had one vote each in the Plenary Assembly.[51]

The specific technical questions were primarily dealt with by the Committees of Rapporteurs. These Committees carried out studies of the issues at

---

[49] Valensi (1929), pp. 3-4.

[50] Jeding (1998), p. 50.

[51] Règlement Intérieur du C.C.I. Téléphonique.

hand, and presented suggestions to the Plenary Assembly who could then accept, modify or reject the Committees' proposals. Issues that were not prepared as draft recommendations by a Committee could not be put on the agenda of the Plenary Assembly. The Plenary Assembly elected which Administrations or companies that were to form the Committees of Rapporteurs, and these delegations designated the individual members of the Committees.[52]

The General Secretary was responsible for co-ordinating the Committees and preparing the following Plenary Assembly. The General Secretary also drew up the agenda of the meetings of the Plenary Assembly, based on the reports with draft recommendations from the Committees of Rapporteurs. For the sake of impartiality the General Secretary was required not to be in active service of any of the member organizations.[53] From the inception of the CCIF until after the Second World War however, Georges Valensi of the French Administration served as General Secretary, and the CCIF's headquarters were located in Paris.

Methodology

In Jeding (1998) the aim is to study how one member state in the CCIF, Great Britain, acted strategically to put forward her national interests to influence the international regulatory system, and how those national interests were formed on the national level.[54] The study draws on Robert Putnam's (1988) concept of the relation between national and international politics as a two-level game. In this model, national positions are formed through the interplay between different interest groups on a national level. On the international level governments try to maximise their own ability to satisfy domestic pressures while minimising adverse consequences of international developments. These games are interrelated, and in order to succeed central decision-makers must pay attention to both levels simultaneously.[55]

The analysis of the British strategic action in this two-level game follows two steps. First a British position is isolated through analysing the various factors which had a direct influence on the international telephone policy of Britain, and the forces which put pressures on it. This includes the study of reports and memoranda from the British Post Office's telephone branch, but also expressed views from other bodies such as the Foreign Office, Treasury, Ministry of Defence, Chambers of Commerce, and others. Since the British policy on international telephony was related to British interests in other

---

[52] Règlement Intérieur du C.C.I. Téléphonique.

[53] Règlement Intérieur du C.C.I. Téléphonique.

[54] Jeding (1998), p. 19.

[55] Putnam (1988).

spheres, it is important to include other interests that may have influenced the formation of a British position.

The second step of the analysis is to then study which strategies Britain used for putting forward her position in the CCIF's work. The CCIF protocols and minutes of the Committees' work together with internal Post Office reports and correspondence give the British side of the CCIF negotiations. Through those sources it is possible to see patterns of behaviour and examples of various strategies used to promote the British national position.

*

In Jeding (2000) the object of study is also the CCIF, although the methodology is altogether different. The aim of the article is to find out which actors were central in CCIF. In other words, given that all members had an interest to influence the CCIF with their own preferences, the article analyses which actors had the best opportunities for doing so.

The article uses network theoretical methods to analyse that question, based on the assumption that the most prominent or important actors are usually located in strategic locations in the network.[56] Centrality is a network theoretical indicator of the extent to which a node is connected to other nodes in the network. *Degree* is a relatively simple and robust measure of centrality, which shows the number of nodes directly linked to one particular node.[57] In this context the participation of individuals, organizations, and countries in the CCIF's Plenary Assemblies (PA's) and Committees of Rapporteurs (CR's) are measured as *degree*.

The members' participation in the Plenary Assemblies and Committees are studied as affiliation networks. An affiliation network is a two-mode network, consisting of both a set of actors and a set of 'events', in this case Plenary Assemblies or Committees. In this way, individual actors are linked not directly to each other, but to a Plenary Assembly or a Committee. *Degree* will then be a measure of the number of PA's or CR's an individual member has participated in. When looking at the organizational and national levels, the *degree* of actors will instead be aggregate measures. Thus the degree of a country will be measured as the sum of all the delegates from that country who participated in the PA's or CR's.

In the study, the PA's and the CR's respectively are treated as two separate affiliate networks. For both these networks the centrality of actors is calculated as degree. The data from each of the studied years are then aggregated into two data sets, covering the whole period 1923-39. This allows results that measure centrality over the whole period, as well as on a year-by-year basis. The study also allows comparisons between the two networks,

---

[56] See for instance Wasserman & Faust (1994), p. 169; Knoke & Kuklinski (1982), p. 52.

[57] Ottosson (1993), p. 26.

in order to analyse whether the actors that were central in one of the networks were also central in the other. The ability to compare centrality of actors in two different networks, on three different levels of aggregation makes it possible to give a relatively fine grained answer to the question of which actors were the most central in the CCIF between 1923 and 1939.

*

In Jeding, Ottosson & Magnusson (1999) the aim is to study the evolution of co-operation between the Scandinavian telecommunication authorities. The study follows this development from the first official meeting of the Scandinavian telegraph Administrations in 1916 up until 1960. This is done by investigating the type of relations between the authorities, and the types of issues that were subject to their co-operation. The study of those parameters allows an analysis of how the co-operation gradually came to include new forms of co-operation, from an originally very limited set of specific issues into a far-reaching programme of joint financial liabilities and joint representation in international bodies. The article also specifically studies the individuals who took part in the meetings between the Administrations and correspondence between them. That gives some ground for reasonable conclusions about the evolving social relations between key functionaries in the Administrations, which helped to further their co-operation.[58]

*

Jeding (2001) studies the shift of the Swedish telecommunications sector from a system of *de facto* monopoly where the State could control the sector through its telecommunications Administration, into a liberalised system; and the change of regulatory regime that accompanied it. The subject of the article is the forms of state intervention and/or regulation directed at achieving certain ends in the sector, under the two different regimes.

The comparison sets out from the policy objectives for the telecommunications sector expressed in the 1993 Telecommunications Act. The means for state intervention or regulation are then examined for each of these objectives under the two regulatory systems respectively. This allows for a comparison of available means of intervention under the different regimes. It also gives room for discussion about the amount of regulatory activity related to achieving those policy objectives under the regimes respectively, and about whether or not the liberalisation of Swedish telecommunications was also coupled to deregulation.[59]

---

[58] Jeding, Ottosson & Magnusson (1999).

[59] Jeding (2001).

26

Sources

The primary sources used for studying the CCIF are above all protocols and minutes from its meetings. After each Plenary Assembly, a preliminary version of the protocols were issued to the member states, which then had an opportunity to comment and correct them. They can therefore be assumed to give correct accounts of what was discussed and decided at the meetings. The protocols also give detailed accounts of which individuals took part in which Committees, and the organizations and states they represented. These data are the basis for the network analyses in Jeding (2000). Copies of the protocols are held both in the archives of the Swedish and British national Administrations and in the ITU archives in Geneva. The ITU archives also include the personal archive of Georges Valensi, the General Secretary of the CCIF for the whole of the period studied. This collection contains some of Mr Valensi's correspondence with the national Administrations as well as his notes for writings on the history of the CCIF.

Documents regarding British strategies in the international telecommunication agreements are found in the archives of the British Post Office's telephone branch, now at BT Archives. It is an extensive and well-ordered archive, containing internal reports and memoranda as well as correspondence. In many cases both sides of the correspondence are kept together, making it easy to follow issues over time. Records from the Foreign Office concerning telecommunications matters are held at the Public Records Office in Kew. In that collection, for instance, reports are held on telecommunications issues and instructions to British delegations at international conferences.

Records of the Swedish participation in international agreements are found in the archives of Telegrafstyrelsen, formerly held at Telia's archives but now at Landsarkivet, Uppsala. This, too, is an extensive archive containing internal reports, memoranda, and correspondence, both regarding the CCIF and direct contacts with other Administrations. Also here are both sides of the correspondence often saved, which gives increased reliability and makes issues easier to follow.

Results

The following section gives a brief summary of the studies in this thesis, and their primary results and conclusions.

National Politics and International Agreements: British Strategies in Regulating European Telephony, 1923-39

When the European national telephone Administrations started co-operating through the CCIF they quickly obtained good results. In the first few years of the CCIF's existence a large number of new telephone connections were opened across the borders of Europe. The described international agreements on technological and administrative standards were however far from any ideal image of straightforward technical issues. Instead the member states held national positions, i.e. sets of preferences with which they tried to influence the whole of the international system for their own benefit. Member states had different national systems, different geographical locations, different needs for international communications etc., so naturally they also had different preferences for the common international standards. This view of the international co-operation as a conflictual process is, however almost completely absent from earlier studies of the CCIF.[60]

The article investigates further how member states acted strategically to achieve their own ends by concentrating on one of them, Great Britain. The whole game of reaching an international agreement is seen as a two-level game. On one level different actors are trying to influence what is to become the national set of preferences – the national position. On an international level this national position is then put forward against the positions of other states. On both levels, actors try to maximise the importance of their relative strengths and minimise the importance of their weaknesses through strategic action.[61]

First, the formation of a British position is studied. It is found that what was to constitute the British position was defined in a complex system of interdependent actors with different preferences on the national level. The outcome of this game was by no means a purely technical standpoint. Instead what was found to be the British position was firmly bound to other political issues.

A number of different factors influenced what came to be the British position. The most important actors on the British national level were the Post Office (which also came to handle the national telephone system after it had been nationalised in 1912); other Government branches such as the Foreign Office and Treasury; industry interests representing radio, telegraph and telephone industry, each with their own set of preferences; user groups, such as international chambers of commerce; and branch specialist organizations, such as the Institution of Electrical Engineers.

Relations between these actors, as well as external factors shaped the formation of the British position. For instance, the relation between the Treasury and the Post Office's telephone branch was one of distrust and conflicting goals. That taken together with the relatively primitive state of the British national system led the British position to be one of insisting on business-like conditions of operation for the telephone system, and a reluctance to devote public finances to investments in telephony. The British po-

---

[60] Jeding (1998), pp. 13-18.
[61] Jeding (1998), pp. 9-13.

sition was also characterised by what might be called a free market policy, which favoured a flexible system of rates and as far as possible competition within the international European system. Historical ties to the Commonwealth, and the need for a global rather than European outlook in matters of foreign policy, made the British position stand out from that of other member states in the CCIF as being far less Eurocentric.[62]

On the national level the formation of a British position illustrates the interconnection between politics and technology in two ways. First there is a clear connection between, for example, foreign policy and British actions in the CCIF. For instance the British demand for extra-European communications was greater than that of most of the other European states, and this in turn led to diverging positions on a number of more technological issues. Secondly the case of establishing a British position gives a good example of how actors with diverging interests were brought into 'antagonistic cooperation'. Through the interplay of different interests, a single British position was negotiated. In this process, issues in the political sphere influenced the relative strength of the actors.

Having established what constituted the British set of preferences the paper then goes on to study which strategies Britain used in order to further her interests in the CCIF. One of these strategies was to try to shift the focus away from cable telephony. Although long-distance telephone technology had improved considerably from its earliest stages, cable telephony was still confined to one continent. Since Britain demanded communications on a global rather than European scale, it was a natural step to promote the development of telegraphy and radio communication technology.[63]

Another British strategy was trying to shift the focus away from Europe. This also had to do with the British demand of global communications, but also with the close links between the British national Administration and the American Bell system. The Institution of Electrical Engineers was a British association for electrical engineering, where representatives from the American Bell company were very active and held a lot of prestige. This further strengthened the ties between the British and American systems. As a consequence, Britain as a rule favoured global rather than European standards in the CCIF. Given that the American system was regarded as the most advanced internationally, this also normally meant that Britain advocated the adoption of American practices as an international standard.[64]

As mentioned, the British position also included the introduction of competition into the international system. Instead of establishing a fixed route for international calls, Britain argued that countries should compete for transit traffic. In this, as in many other cases, British interests were opposed to

---

[62] Jeding (1998), pp. 75-83.

[63] Jeding (1998), pp. 86-90.

[64] Jeding (1998), pp. 90-92.

those of Germany. To some extent ideological reasons can be cited, but also to no small extent the explanation can be sought in the geographical location of the countries. Britain on the periphery of Europe had almost no transit traffic running through its territory at all. Germany on the other hand, located in the centre of the continent, had a powerful position as a transit country, and had no desire to face competition for its transit incomes.[65]

A fourth strategy was to influence the CCIF through expertise. This meant that Britain devoted considerable resources to supplying experts to the various committees of the CCIF. The British Administration seems to have regarded this as a useful strategy for influencing the European system with its own preferences. It was also a strategy where Britain could make the most of the fact that there were many highly skilled electrical engineers that were British. Through the Institution of Electrical Engineers, these British engineers were also familiar with developments in the American system.[66] As a fifth and last strategy, Britain could try to block the reaching of a decision in the CCIF. What is most surprising about this strategy is that it worked at all. Formally each member state had one vote in the Plenary Assembly. Informally however, there was a strong tendency towards reaching consensus solutions, and this could occasionally be used to block decision making in those cases where all other strategies had failed. This was probably only useful as a seldom-used strategy. Had Britain tried to use it more frequently, it would most likely have been ignored by the other member states.

Many of these strategies built on the greater demand of extra-European communications of Britain than of other European states. These strategies also filled the function of diminishing the disadvantage of being geographically peripheral in Europe. Apart from those strategies, Britain also tried to make the most of her good supply of highly skilled engineers. This was done through devoting resources and expertise to the expert study groups of the CCIF. The common features of all this strategic action is that actors, on the national as well as the international level, used strategies that made the most of their relative strengths and diminished the importance of their weaknesses.

The main contribution of this paper is to highlight the conflictual side of the international telecommunications agreements, and providing an instrument for analysing the behaviour of national representatives in terms of national politics rather than some isolated technical efficiency.

Networks of Telephony: Central Actors in the CCIF, 1923-39

In Jeding (2000) member states' strategic action in the CCIF is studied further. The aim of the article is to find out which actors were central in the

---

[65] Jeding (1998), pp. 93-97.

[66] Jeding (1998), pp. 97-100. This point is taken further in Jeding (2000).

CCIF's Plenary Assemblies and Committees of Rapporteurs.[67] The concept of centrality is fetched from network theory, and used as an indicator of which actors had the highest number of contacts with other actors in the CCIF network, through participating in Plenary Assemblies (PA's) and/or Committees of Rapporteurs (CR's).

In the article, the PA and the CR's are treated as separate affiliate networks. For both of these networks the centrality of the participating members is calculated, measured as *degree*. These calculations are then aggregated to the organization the individual members represent, and the state they represent. In that way centrality of the actors is measured on three different levels; individuals, organizations, and states, for both networks.[68]

The underlying assumption of the study is that actors that participated in a PA or CR had better opportunities for influencing the CCIF's decisions than those actors that did not participate. Further it is assumed that an organization or a country that had a large number of representatives in the PA's or CR's had an advantage over an organization or country with fewer representatives, all other things equal, in influencing the CCIF's decision making.[69]

One immediately visible feature of the CCIF networks is that the number of participants increased over the studied period. 19 members attended the first meeting in 1923. Over time the conferences grew more complex, with the number of participants reaching 139 in 1936.[70]

One of the aims of the article was to find out which states were dominant in the networks over the period. The answer is that Great Britain with some margin was the most central state, both in the PA and CR networks. After Great Britain there was a gap before France and Germany in second and third place. The relative position of those two changed over the period, though, with Germany being more central than France during the first half of the period. After those three countries there was a group of Western and Central European states that together made up a core of central states in the CCIF. It is however worth noting that over the period the centrality of non-European countries increased.[71]

There were some differences as to which countries were central in the Plenary Assemblies and those that were central in the Committees of Rapporteurs, but on the whole there was a remarkably strong correlation between the two networks. This leads to the conclusion that those states that were important in the CCIF tended to be well represented in both those bodies, although exceptions to that rule can be found.

---

[67] The structure of the CCIF and the function of these bodies are briefly described above.

[68] Jeding (2000), pp. 6-7.

[69] Jeding (2000), p. 7.

[70] Jeding (2000), p. 12.

[71] Jeding (2000), pp. 22-23.

Another general conclusion is that the most central organizations tended to represent the most central countries. The British Post Office was by far the most central organization, followed by the French Ministry of Posts, Telegraphs, and Telephones, and the German Ministry of Posts in second and third place respectively. Similarly to the national level there was a dominance of organizations representing Western and Central European states. The two main exceptions to that rule were the American long-distance operator AT&T and the Soviet People's Commissariat of Posts, Telegraphs, and Telephones. There was also a strong correlation between the centrality measures for organizations in the two networks. In other words, those organizations that were well represented in the PA's tended to be so also in the CR's.[72]

Another striking fact about the most central organizations is that most of them were public telephone Administrations, in most cases with national monopolies. Of the 15 most central organizations in the Plenary Assemblies, the American Telephone and Telegraph Company was the only private operator, albeit with a national monopoly on international traffic. This is a result that must be taken into account when discussing the type of recommendations that came from the CCIF. The fact that all the most influential players in the CCIF were monopolists, most of them public, surely can say something about the persistence of central regulation in the field of telephony, and about the long standing view of telephony as a natural monopoly.[73]

Centrality at the individual level gives a less clear-cut picture than the organizational and national levels. There was a strong correlation between the most central individuals in both networks. In other words, those individuals who were members of a large number of Committees were generally also participating in a large number of Plenary Assemblies over the period. The list of most central individuals did not however follow the list of most central organizations or states as closely. The pattern with the most central members representing the core of Western or Central European states is however repeated here as well.[74]

On the whole, the results in the article show that Great Britain devoted relatively large resources towards participating in and influencing the CCIF's work. France and Germany were also key actors over the period, although not as central as Britain. Given that member states had an interest in influencing the emerging common institutions for international telephony with their own preferences, Britain, and to some lesser extent, France and Germany had an advantage in pursuing this. The results also confirm the image of the inter-War CCIF as a rather introvert European organization,

---

[72] Jeding (2000), p. 23.

[73] Jeding (2000), pp. 23-24.

[74] Jeding (2000), pp. 23-24.

dominated by and primarily interested in the European telephone system.[75] After the Second World War, CCIF and ITU were thoroughly reorganized, brought into the UN system, and turned into more truly international bodies. The results in the article however points strongly to the conclusion that before the War, Western and Central European states had a strong advantage in setting the rules for the international telephone system.[76]

Regulatory Change and International Co-operation: The Scandinavian Telecommunication Agreements, 1900-1960

The third article is also concerned with international agreements in telecommunications. Here however the focus is on the voluntary, often informal bilateral or multilateral agreements that take place outside formal organizations. The empirical focus is on telecommunication agreements between the Scandinavian countries during the period 1900-1960.

The origins of the Scandinavian co-operation were a conference concerning censorship of telegraph traffic. In 1916 representatives from the telegraph and telephone Administrations of Denmark, Norway, and Sweden met in Copenhagen. Their purpose was to find common and more effective rules of censorship, to hinder the transmission of information that could harm the merchant ships of the neutral Scandinavian countries during the War.

Following this first meeting between the Administrations, they gradually came to meet more often. The meetings were originally initiated by the respective ministries of foreign affairs, and their purpose was to deal with more specifically political questions. Over time, though, the Administrations saw the uses of such international co-operation and started meeting on their own initiative.

Gradually the scope for co-operation expanded so that the Administrations came to work together on a number of different fields. One such issue was line construction. During the First World War the demand for international telephone calls in Scandinavia increased dramatically. Due to wartime materials shortages, line construction was incapable of keeping up with demand. Express calls made up half of all the calls made.[77] The majority of international calls from any of the Scandinavian countries terminated in one of the other Nordic countries, and those that did not typically had to be connected through another Scandinavian transit country. This close interconnection led the Scandinavian Administrations to co-operate over the construction of new telephone lines.[78]

---

[75] Cf. Schmidt & Werle (1999), p. 30.

[76] Jeding (2000), p. 24.

[77] Jeding, Ottosson, & Magnusson (1999), pp. 69-70.

[78] Heimbürger (1968), p. 40; Jeding, Ottosson, & Magnusson (1999), p. 71.

The co-operation between the Scandinavian Administrations was taken one step further when the first Trans-Atlantic telephone cable was constructed. That cable was owned jointly by the American long-distance telephone operator AT&T, the Canadian Overseas Telecommunication Corporation, and the British Post Office. The Scandinavian states were jointly offered one channel in that cable against a guaranteed minimum transit fee to the British Post Office. Similar agreements were made later as new cables were laid. The Scandinavian authorities also went into shared economic liabilities as they bought rights of use of a British satellite station.[79]

The established links between the Scandinavian Administrations were further strengthened, as international co-operation in matters of telephony became more firmly institutionalised. As the CCIF was formed to deal with the problems of a fragmented telephone system in Europe, the Scandinavian representatives very early realised that they had common interests in this forum. Already in 1924, the Swedish and Norwegian Administrations joined forces to make sure that Germany would be taken into the European co-operation.[80] In 1925 the Swedish, Norwegian, and Danish Administrations again prepared a common position in relation to a question in the CCIF, and were successful in achieving their goal.[81] Following these early successes, the Scandinavian Administrations have subsequently often acted as a bloc in the CCIF and other international organizations. On a number of occasions they have even gone so far as to send one or more common representatives to international meetings to represent their identified common positions.[82]

As illustrated above, the Scandinavian co-operation gradually expanded from its early, very focused field of interest into more diverse forms where a common interest has been identified. Interestingly, these new areas of co-operation were dealt with through the same organizational form that was built up earlier and for different purposes. That suggests that the form for the co-operation was path dependent. Once a successful form for dealing with one type of issues had been established it was used also for other types of questions. That could suggest that the very act of co-operation created channels of contact and social networks between the organizations, which could then be drawn upon to establish co-operation in new fields as well.

These established relations between the Scandinavian Administrations could then be used to their advantage in more formal international organizations, such as the CCIF. This development indicates support for the historical institutionalist idea that not only do institutions put bounds within which self interested actors can act freely - the institutions and historical experience of actors also seem to influence their preferences.

---

[79] Jeding, Ottosson, & Magnusson (1999), pp. 71-72.

[80] Jeding, Ottosson, & Magnusson (1999), pp. 73-74.

[81] Jeding, Ottosson, & Magnusson (1999), p. 74.

[82] Jeding, Ottosson, & Magnusson (1999), p. 74.

Liberalisation and Control: Instruments and Strategies in the Regulatory Reform of Swedish Telecommunications

In the fourth article the focus is on national decision making in the telecommunications sector. As on the international arena, the national telecommunications systems need co-ordination in order to work coherently. But in addition to those functional needs for co-ordination, the State typically sets up different policy objectives for the sector. These are different kinds of effects that the State finds desirable, and therefore want to direct the sector so as to achieve them.

One important difference between the international and the national arena is that on the national level there is a hierarchically superior actor, the State, that can impose its rules on the other players. However, those rules have to be matched to the overall structure of the sector. When the industrial structure of the sector changes, the institutional structure has to change as well, in order for the rules to be effective.

When the Swedish Parliament in 1992 decided that the Swedish telecommunications market should be opened to competition, this meant that the State could no longer control the whole sector directly through the telecommunications Administration, Televerket. The policy objectives however remained the same. This created what has been called a 'capability gap' between governmental goals and capabilities.[83] The aim of the article is to study how the State acted to close that gap through a regulatory reform of the telecommunications sector.

Throughout most of the 20th century, telecommunications were among the most strictly regulated areas of most of the national economies in Europe. In Sweden the State control over the sector was executed through the national telecommunications Administration Televerket. It was established as a State-owned public enterprise[84]. That kind of enterprise has traditionally used in Swedish administration in the communications sector and in some other industries regarded as vital and/or strategic for the nation. Organizationally it takes a middle ground in the administrative system. On the one hand it is more independent from the State than a regular civil-service authority, both relating to its goals and how these are achieved. On the other hand it is considerably more restricted in its actions than a State-owned limited liability company.[85]

Through the direct control over Televerket, the Government could use it as an instrument for its telecommunications policy. Televerket was for instance not an independent legal subject. The Administration had to follow

---

[83] See Vogel (1996), p. 25.

[84] The term in Swedish is 'affärsverk'. Karlsson (1998) translates this as 'State-owned public enterprise', whereas Noam (1992) uses the term 'public service corporation'.

[85] Karlsson (1998), p. 79.

administrative laws and regulations like a civil-service authority, and its decisions could be appealed against to the Government. All its major and strategically far-reaching decisions had to be taken by the Government or Parliament, and the Parliament also decided on its annual budgets and investment plans. Besides this, the Parliament could also set the prices of Televerket's products and services, and its assets were part of State property. This all meant that Televerket's Director-General was primarily responsible to the Government rather than to the board. What gave Televerket some autonomy over the telecommunications policy was the asymmetric relationship between Televerket and the Ministry for Communications, where the latter was very small and almost all the staff and expertise in telecommunications matters was with the Administration.[86]

In 1993 the Swedish Parliament adopted a new telecommunications legislation bill. Its purpose was to open the Swedish market to competition, and it included a new Telecommunications Act, a revised Radiocommunications Act, and a proposal to corporatise Televerket. In the bill, the State's policy objectives for the telecommunications sector were revised. The State's responsibility for ensuring universal service to all subscribers regardless of geographical location was expanded. Also telefax and low speed modem data communication should now be included in the universal service requirements.[87]

Most importantly however, a new policy objective was to create opportunities for an efficient competition in the telecommunications markets. Competition should be the instrument with which consumers would get a wider choice of services at higher quality and lower prices.[88]

One of the most important implications of the introduction of competition on the market and the corporatisation of Televerket was that the new corporation Telia had to start acting as an independent operator. Telia could no longer act as the State's instrument for telecommunications policy, and a new regulative body was created to perform that function. The National Post and Telecom Agency (PTS) was created as an independent regulator for the sector. PTS was installed as the Government's 'watchdog' with an overall responsibility for the telecommunications sector. It took over a number of authorities from Televerket, such as responsibility for the national numbering plan, frequency administration, standardisation issues, and representing Sweden in international co-operation on telecommunications.[89]

A new regulatory task of PTS was to ensure that efficient competition was established and upheld in the various segments of the telecommunications market. Among the tasks associated with this goal were issuing li-

---

[86] Karlsson (1998), pp. 79-80; Ioannidis (1998), ch. 7.

[87] Jeding (2001), p. 5.

[88] Telecommunications Act 1993:597, s. 3.

[89] SOU 1992:70, ch. 7.

censes to operators, mediating between operators and even make binding decisions in cases where operators could not agree on agreements over e.g. interconnection fees, and generally upholding the sector specific competition rules.[90]

A comparison of the two regulatory systems shows that the policy objectives, as expressed in the Telecommunications Act, remained stable over the regulatory reform. The objectives of the State were primarily to ensure efficient telecommunications, sustainable and accessible services during crises and wartime, and universal services for all subscribers at reasonable prices. In addition to these objectives, the new goal of ensuring efficient competition was added.

The liberalised system demanded that Telia should act as an independent operator, and its goals should be strictly based on business considerations. The state thus had to find another instrument for realising its policy objectives. Where telecommunications policy under the old system could be pursued through directives to Televerket, the liberalised system called for neutral regulations. Thus PTS became the instrument for State intervention in the sector, with a mandate to enforce regulations and condition licenses to operators.

Through the regulatory reform the State relinquished its direct control over the telecommunication sector through directing Televerket. The new liberalised system with the independent regulator PTS as its instrument demanded more general, neutral regulations applicable to all operators. Regulation also had to shift from ex post to ex ante. The comparison of regulation under the two systems further shows that liberalisation in fact did not mean deregulation, in the sense that the amount of regulations or regulatory activity did not decrease. Instead, a number of new regulatory tasks were added with the new policy objective of ensuring efficient competition.[91]

---

[90] Jeding (2001), pp. 5-10.
[91] Jeding (2001), pp. 10-12.

Conclusions

Telecommunications is one clear example of a network industry that requires an extensive co-ordination in order for its parts to work together as an integrated system. This thesis studies three different cases of how actors have tried to achieve such co-ordination. One method was to create an international organization with the task of finding common rules for the interconnection of national systems. A second method was through reaching voluntary, non-binding bilateral agreements, outside any formal organization. The third example is when, on the national arena, the state imposed new rules of the game for the industry.

The first alternative solution to the co-ordination problem created a relatively stable institutional environment for the participating actors. One effect of this was to establish a mechanism for problem solving and collective decision-making, which resulted in rapid progress in the interconnection of the national telephone systems in Europe. Another effect was that the fixed institutional order of the organization CCIF also meant a stable mechanism for member states that wanted to exert their influence on the organization's decisions. The institutional arrangements thus defined which type of behaviour was profitable for states and organizations that wanted to influence the CCIF's decisions.

In an international setting, the co-ordination of national telecommunications systems was a process that took place between sovereign states. In the case of the CCIF participation was voluntary, as was the adherence to its rules. International co-ordination had to occur without any party having the power to enforce its will on the others. Nevertheless, the recommendations issued by the CCIF were almost universally followed by the member states. Moreover, states that had an interest in influencing the CCIF's recommendations could gain positions of influence through supplying expertise and participating extensively in the CCIF's committees.

An analysis of the centrality of states in the CCIF between the years 1923-1939 shows that during this period the organization was dominated by a core of Western and Central European states and by public monopoly telephone administrations. It is reasonable to assume that this may have influenced the CCIF's recommendations concerning central regulation of telephony and the view of telephony as a natural monopoly. Towards the end of the period however, other types of actors from other parts of the world gained influence.

Regarding the voluntary co-operation outside formal organizations that took place between the Scandinavian telecommunications administrations another pattern is visible. These administrations were originally brought together for very specific tasks, but once a channel for co-operation had been opened between them their fields of co-operation expanded. Thus what originally was limited to very specific issues gradually broadened into a very far-reaching co-operation, such as common representation in international bodies and joint economic liabilities. This might in part be explained by the evolution of a social network between the actors and organizations, as the individual participants in the Scandinavian meetings showed a very stable pattern over time.

For both of the above mechanisms for co-ordination, a pattern of institutional development is visible. Both the CCIF and the more loosely organised Scandinavian co-operation started relatively modestly with a rather narrowly defined agenda. Over time both the scale and scope of the co-operation were expanded and intensified whereas the institutional framework governing that work remained relatively stable. This indicates some amount of path dependency in the institutional development: once a form was found for handling the problems of reaching collective decisions despite the different preferences of the actors, this form was then applied to new problems that arose.

Another observation is that in both these cases the whole process of international co-ordination and co-operation, once established, seemed to reinforce itself and gather a momentum of its own. This finding is in line with the findings e.g. in Krueger (1996), who explains it with the fact that complexity of regulatory programmes in itself makes the specialists in those programmes a vested interest in the maintenance of the programmes. Thus in the case of the co-ordination of the international telecommunications system, the expansion and complexity of a fixed form for reaching agreements provided a barrier of entry to that work of non-specialist groups.

*

On the national level the regulation process was of a different kind. Here, the state held a hierarchically superior position, which allowed it to impose its rules on the national arena. However, when the Swedish parliament decided to liberalise the telecommunications sector, this implied that the state relinquished its tools for direct control of the sector. A liberalised market was not compatible with direct control over the monopoly operator. Instead a new regulatory system was implemented, which relied on more general legislation and indirect, *ex ante* rules. An independent regulatory agency was also created, to act as a neutral 'watchdog' over the liberalised sector. A whole range of new instruments for state co-ordination and control was added, and the state's need for control over the sector remained stable or even increased. This gave the effect that a freer market demanded more rules.

In other words, on the national level the state had the formal power to impose its will on all the other actors, and thus there was no need for an elaborate system for reaching decisions regarding the telecommunications sector. Despite this the liberalisation of the Swedish telecommunications sector led to the creation of a large set of new rules. The need for a new regulatory function was coupled to the political objectives for the sector defined by the state, and with the creation of a liberalised system came the demand for a whole system of neutral, indirect rules pertaining to all the actors on the market.

*

It is wholly insufficient to simply regard the regulation and co-ordination of telecommunications as a technical response to a technical need. Many of the earlier studies of this kind of activities, especially on an international level, regard them as purely technical exercises. This thesis shows that the choice of common international standards and practices in the inter-War period was closely linked to other national and political interests. The rhetoric from the participants in the international meetings describes their work as a harmonious co-operation to find technically optimal solutions. Despite this, the actions of member states suggest that they had strategic interests in influencing the outcome of the negotiations with their own national preferences. Furthermore, these strategic interests were often based on other policy objectives than the purely technical.

References

Andersson-Skog, L. (1999), 'Political Economy and Institutional Diffusion: The Case of Swedish Railways and Telecommunications up to 1950', in Andersson-Skog, L. & Krantz, O. (eds.), *Institutions in the Transport and Communications Industries*, Science History Publications, Canton, Mass.

Baldwin, R. & Cave, M. (1999*), Understanding Regulation: Theory, Strategy, and Practice*, Oxford University Press, Oxford.

Berg, A. (1999), *Staten som kapitalist. Marknadsanpassningen av de affärsdrivande verken 1976-1994*, Uppsala University Library, Uppsala.

Bobroff, S. A. (1974), *United States Treaties and other International Agreements Pertaining to Telecommunications*, U.S. Department of Commerce/Office of Telecommunications.

Chapuis, R. (1976), 'The CCIF and the Development of International Telephony, 1923-56', *Telecommunication Journal, 43:III*, pp. 184-197.

Clarke, A. C. (1958), *Voice Across the Sea*, Harper & Brothers, New York.

Codding Jr, G. A. (1952), *The International Telecommunication Union: An Experiment in International Cooperation*, E. J. Brill, Leyden.

De Bijl, P. (2000), *Competition and Regulation in Telecommunications Markets*, CPB Netherlands Bureau for Economic Policy Analysis, The Hague.

Dixit, A. K. (1996), *The Making of Economic Policy – A Transaction-Cost Politics Perspective*, MIT Press, Cambridge, Mass.

Ds 1996:29, *Nästa steg i telepolitiken*, Kommunikationsdepartementet, Stockholm.

Ds 1996:38, *Moderna telekommunikationer åt alla*, Kommunikationsdepartementet, Stockholm.

Foreman-Peck, J. & Millward, R. (1994), Public and Private Ownership of British Industry, Clarendon Press, Oxford.

Fossion, H. (1938), 'Le comité consultatif international téléphonique – son origine: son évolution', *Journal des Télécommunications*, 5:XII, pp. 337-343.

Francis, J. (1993), *The Politics of Regulation: A Comparative Perspective*, Blackwell Publishers, Oxford -Institut für Gesellschaftsforschung, Cologne.

Goldin, C. & Libecap, G. (eds.) (1994), *The Regulated Economy: A Historical Approach to Political Economy*, University of Chicago Press, Chicago.

Grimm, K. D. (1972), *The International Regulation of Telecommunication, 1865-1965*, University of Tennessee, UMI Dissertation Services, Ann Arbor.

Headrick, D. R. (1981), *The Tools of Empire*, Oxford University Press, Oxford.

Headrick, D. R. (1991), *The Invisible Weapon: Telecommunications and International Politics, 1851-1945*, Oxford University Press, Oxford.

Heimbürger, H. (1953), *Svenska Telegrafverket IV*, Televerket, Stockholm.

Heimbürger, H. (1968), *Nordiskt samarbete på telekommunikationsområdet under 50 år, 1917-1967*, Televerket, Stockholm.

Heimbürger, H. (1974), *Svenska Telegrafverket V:I*, Televerket, Stockholm.

Helgesson, C.-F. (1995), 'Technological Momentum and the "Natural" Monopoly', *Paper presented at the SHOT 1995 annual meeting.*

Helgesson, C.-F. (1999), *Making a Natural Monopoly: The Configuration of a Techno-Economic Order in Swedish Telecommunications*, EFI: The Economic Research Institute, Stockholm School of Economics, Stockholm.

Hodgson, G. M. (1988), *Economics and Institutions: A Manifesto for Modern Institutional Economics*, Polity Press, Cambridge.

Hughes, T. P. (1987), 'The Evolution of Large Technological Systems', in W. E. Bijker, T. P. Hughes and T. J. Pinch (eds.), *The Social Construction of Technological Systems: New Directions in the Sociology and History of Technology*, MIT Press, Cambridge, Mass.

Immergut, E. M. (1992), 'The rules of the game: The logic of health policy-making in France, Switzerland, and Sweden', in Steinmo, S., Thelen, K., and Longstreth, F., (eds.) *Structuring Politics*, Cambridge University Press, Cambridge.

Ioannidis, D. (1998), *I nationens tjänst? Strategisk handling i politisk miljö*, EFI: The Economic Research Institute, Stockholm School of Economics, Stockholm.

Israelsen, H. (ed.) (1992), *P&Ts historie, band 3: 1850-1927*, Generaldirektoratet for Post og Telegrafvæsenet, Copenhagen.

Jeding, C. (1998), 'National Politics and International Agreements: British Strategies in Regulating European Telephony, 1923-39', *Working Papers in Transport and Communication History 1998:1*, Departments of Economic History, Umeå University and Uppsala University, Uppsala.

Jeding, C. (2000), 'Networks of Telephony: Central Actors in the CCIF, 1923-39', *article presented at the 2000 Communications Policy Workshop, Uppsala.*

Jeding, C. (2001), 'Liberalisation and Control: Instruments and Strategies in the Regulatory Reform of Swedish Telecommunications', in Magnusson, L. & Ottosson, J. (eds.) *Interest Groups and the State,* Edward Elgar, Cheltenham.

Jeding, C., Ottosson, J. & Magnusson, L. (1999), 'Regulatory Change and International Co-operation: The Scandinavian Telecommunication Agreements, 1900-1960', *Scandinavian Economic History Review,* 47:2, pp. 63-77.

Kaijser, A. (1994), *I fädrens spår...*, Carlssons, Stockholm.

Kaijser, A. (1999), 'The Helping Hand: In Search of a Swedish Institutional Regime for Infrastructural Systems', in Andersson-Skog, L. & Krantz, O. (eds.), *Institutions in the Transport and Communications Industries*, Science History Publications, Canton, Mass.

Karlsson, M. (1998), *The Liberalisation of Telecommunications in Sweden*, Department of Technology and Social Change – Tema T, Linköping University, Linköping.

Karlsson, M. & Sturesson, L. (eds.) (1995), *Världens största maskin*, Carlssons, Stockholm.

Keohane, R. O. (1999), 'Ideology and Professionalism in International Institutions: Insights from the work of Douglass C. North', in Alt, J. E., Levi, M. & Ostrom, E. (eds.), *Competition and Cooperation: Conversations with Nobelists about Economics and Political Science*, Russell Sage Foundation, New York.

Knoke, D. & Kuklinski, J. H. (1982), *Network Analysis*, Sage Publications, Newbury Park, Ca.

Krueger, A. O. (1996), 'The Political Economy of Control: American Sugar', in Alston, J., Eggertsson, T. & North, D. (eds.), *Empirical Studies in Institutional Change*, Cambridge University Press, Cambridge, pp. 169-218.

Laffont, J.-J. & Tirole, J. (1993), *A Theory of Incentives in Procurement and Regulation*, MIT Press, Cambridge, Mass.

Laffont, J.-J. & Tirole, J. (2000), *Competition in Telecommunications*, MIT Press, Cambridge, Mass.

Lee, K. (1996), *Global Telecommunications Regulation: A Political Economy Perspective*, Pinter Press, London.

Levy, B. & Spiller, P. T. (eds.) (1996), *Regulations, Institutions, and Commitment. Comparative Studies of Telecommunications*, Cambridge University Press, Cambridge.

Magnusson, L. & Ottosson, J. (2000), 'State Intervention and the Role of History – State and Private Actors in Swedish Network Industries', *Review of Political Economy*, 12:2, pp. 191-201.

Mölleryd, B. (1999), *Entrepreneurship in Technological Systems – The Development of Mobile Telephony in Sweden*, EFI: The Economic Research Institute, Stockholm School of Economics, Stockholm.

Noam, E. (1992), *Telecommunications in Europe*, Oxford University Press, Oxford.

North, D. C. (1990), *Institutions, Institutional Change, and Economic Performance*, Cambridge University Press, Cambridge.

Ottosson, J. (1993), *Stabilitet och förändring i personliga nätverk: Gemensamma styrelseledamöter i bank och näringsliv 1903-1939*, Uppsala University, Uppsala.

Peltzman, S. (1989), 'The Economic Theory of Regulation after a Decade of Deregulation', *Brookings Papers on Economic Activity: Microeconomics*, pp. 1-59.

Perrow, C. (1984), *Normal Accidents: Living with High-Risk Technologies*, Basic Books, New York.

Pitt, D. (1980), *The Telecommunications Function in the British Post Office: A Case Study of Bureaucratic Adaption*, Saxon House, Teakfield Ltd., Westmead, Farnborough.

Putnam, R. (1988), 'Diplomacy and Domestic Politics: The Logic of Two-level Games', *International Organization*, 42:3, pp. 427-460.

Rafto, T. (1955), *Telegrafverkets historie, 1855-1955*, John Griegs Boktrykkeri, Bergen.

Règlement Intérieur du C.C.I. Téléphonique, *ITU Archives: Depot de M. Valensi.*

Riksdagens revisorer (1995), *Uppföljning av post- och telepolitiska mål*, 1995/96:5, Stockholm.

Riksrevisionsverket (1995), *Två år med telelagen*, 1995:31, RRV, Stockholm.

Sampson, A. (1973), *The Sovereign State: The Secret History of the ITT*, Hodder and Stoughton, London.

Schmidt, S. K. & Werle, R. (1998), *Coordinating Technology: Studies in the International Standardisation of Telecommunications*, MIT Press, Cambridge, Mass.

Schneider, V. (1991), 'The Governance of Large Technical Systems: The Case of Telecommunications' in T. La Porte (ed.) *Social Responses to Large Technical Systems*, Kluwer, Amsterdam.

Skårfors, R. (1997), Telegrafverkets inköp av enskilda telefonnät. Omstruktureringen av det svenska telefonsystemet 1883-1918. *Working Papers in Transport and Communication History 1997:3*, Departments of Economic History, Umeå University and Uppsala University, Uppsala.

Sobel, R. (1982), *ITT: The Management of Opportunity*, Sidgwick & Jackson, London.

43

Stigler, G. J. (1971), 'The Theory of Economic Regulation', *Bell Journal of Economics and Management Science, 2 (1), Spring*, pp. 3-21.

SOU 1992:70, *Telelag. Betänkande av telelagsutredningen.* (Report from the Government's Commission on a new Telecommunications Act), Kommunikationsdepartementet, Stockholm.

Telecommunications Act, SFS 1993:597.

Tullock, G. (1994), *Den politiska marknaden – Introduktion till public choiceskolan*, Timbro, Stockholm.

Valensi, G. (1929), *The First Five Years of the International Advisory Committee for Long-Distance Telephone Communications*, Verlag Europäischer Fernsprechdienst, Berlin, ITU Archives.

Valensi, G. (1956), *Le Comité Consultatif International Téléphonique (C.C.I.F.) 1924-1956*, Unprinted manuscript, ITU Library.

Valensi, G. (1965), 'The Development of International Telephony: The Story of the International Telephone Consultative Committee (CCIF) 1924-1956', *Telecommunications Journal 32:I*, pp. 9-17.

Vogel, S. K. (1996), *Freer Markets, More Rules*, Cornell University Press, New York.

Wasserman, S. & Faust, C. (1994), *Social Network Analysis: Methods and Applications*, Cambridge University Press, Cambridge.

National Politics and International Agreements.

British Strategies in Regulating European Telephony, 1923-39.

(Published in *Working Papers in Transport and Communication History 1998:1*, Departments of Economic History, Umeå University and Uppsala University, Uppsala, 1998.)

Innehåll

Acknowledgements

This licentiate thesis was presented and accepted for the degree of *Filosofie licentiat* on February 16, 1998. Throughout the work with it I have gradually come to owe plenty of people a lot of thanks. The organizational setting in which I have worked is the research programme *Communications in Perspective* at the Department of Economic history, Uppsala University. Within that programme I have enjoyed the supervision and support of Dr Jan Ottosson and Professor Lars Magnusson. The programme in its entirety has provided a good environment in which to work, most of all due to the people in it. Apart from the two already mentioned, Dr Juan Bergdahl, Magnus Carlsson, Eva Liljegren, Rikard Skårfors and Jan Östlund have all in their own way helped me with the most diverse aspects of my work. There is also reason to thank the participants at the departmental seminar who, in the early stages of this work, made comments on it which have saved me both time and misdirected effort.

Most of the empirical work presented here was carried out during the academic year 1996/97 when I visited St Antony's College, Oxford; and if ever there was a creative environment, St Antony's is it. During my stay there Dr James Foreman-Peck generously shared his time, expertise and ideas, and for that I am grateful. Special thanks are due to Mark Hickford, whom I seriously tried to wear out with endless discussions over innumerable cups of coffee, but failed. He and Carol Brown also gave their best shot at making my English intelligible. My thanks also go to the participants at the Graduate Workshop in Social and Economic History at Nuffield College, who generously gave of their time and efforts.

Staff at various libraries and archives have of course contributed. First among these are the staff at BT Archives in London, whose help at times have gone beyond the call of duty. Others are the people at the university libraries in Uppsala and Oxford, at Telia's archives in Farsta, the Public Records Office in Kew, and at Landsarkivet in Uppsala.

Finally my gratitude for financial support should be expressed. First of all to the Swedish Transports and Communications Research Board (Kommunikationsforskningsberedningen), and secondly to the British Council who, through the Prince Bertil Scholarship for 1996 made my stay in Oxford a lot easier.

Uppsala, June 1998

Carl Jeding

1. Introduction

The subject matter of this thesis is the development of a system of common rules for the international telephone network in Europe. At some point early in the 20th century, the spread of the telephone had reached a point where subscribers started demanding long-distance lines. Fairly early in the century, the technology of long-distance telephony had also reached a stage where it enabled telephone conversations over great distances. In the United States trans-continental telephone lines were established in the 1910's.

In Europe however, the telephone system was in fact not one, but a number of fairly isolated, national systems. In most cases these systems had developed independently of each other, with the effect that they used somewhat different technology and practices, which created difficulties when the national systems started interconnecting. As a way of overcoming these problems, a number of European states formed an organisation called CCIF in 1924.[1] The purpose of the CCIF was to reach agreements on common standards and practices in the international telephone service, so as to create a coherent European network, where it would be possible to make telephone calls from one end of Europe to another.

From the nature of the problem follows that their solution would have to come through international agreements. Furthermore the fact that the national telephone operators in most cases were state-owned monopolies, and thus to some extent politically influenced, indicates that the co-ordination process was of a political nature. Yet many of the students of international co-operation in the telecommunications sector, both of the CCIF and of the International Telecommunications Union, tend to emphasise the non-political, technical nature of the agreements.

In this thesis I will argue that the co-ordination indeed was a political process, or, more specifically, that the seemingly technical issues had political connotations. If the problem simply was to find common standards and practices, regardless which as long as they were common, there would be no need for strategic action within the

---

[1] The organization is today known as CCITT, and was before 1932 called CCI. For the sake of clarity I will however use the term CCIF consistently.

CCIF. Still it becomes apparent when studying the decisions and decision-making processes within the organisation, that the member states have national 'positions' or objectives which they are trying to pursue. This suggests that the member states' actions within the CCIF are linked to larger political issues, and that the politics of regulating telephony can not fruitfully be studied in isolation from wider aspects of both national and international politics.

In my attempt to study the interrelatedness of these different political fields, I have chosen to study the case of one state, Great Britain, and how the British representatives in the CCIF tried to influence the whole of the European telephone system with their national preferences.

1.1 Network industries and their need for co-ordination

The sector of communication is traditionally a strongly regulated area in the public economy. Many of the markets for transports and communications of different kinds have been considered natural monopolies, wherefore the state has decided that different kinds of regulations would provide better solutions for society as a whole, than the unregulated market. These rules and regulations have been formed considering political and economic goals, but also considering the needs or wishes of concerned actors. Thus the formal rules regulating the interplay on the markets for communications have had a strong effect on the way in which these markets have worked.[2]

Infrastructural industries are in most cases network industries, i.e. their operation is based on a network of nodes of some kind. That makes co-ordination of the different parts of the network important for its function. One theoretical concept dealing with such needs for co-ordination is that of technical linkage.[3] The meaning of this concept is the importance for the network as a whole that its parts can be co-ordinated. In systems with a strong technical linkage between its different components, such as e.g. railways or telephone networks, an extensive co-ordination is needed for the network to function, and extensive vertical integration is therefore common within the network. In systems with a lesser degree of technical linkage,

---

[2] See for instance Foreman-Peck and Millward (1994), ch. 1; Andersson-Skog & Ottosson (1994)

such as airline traffic or even better bus traffic, the occurrence of vertical integration is usually rarer due to the lesser need for co-ordination.[4]

The presence of horizontal integration is also related to the degree of technical linkage. Strongly linked systems have often been organised in the form of monopolies, many times state-owned, while the less linked systems more often have worked as markets open for competition. Again, the need for co-ordination probably has been a factor here, but also the fact that the systems with a strong technical linkage have been regarded as natural monopolies.[5]

Judging from these theories then, telephone systems require extensive co-ordination in order to function satisfactory, and could therefore be expected to show a tendency towards both vertical and horizontal integration. On the national levels this development is clear. In most European countries legal monopolies for operating both local and long-distance telephony were established. On the European level however, no such integration tendencies were recognisable and, indeed, co-ordination of the European system as a whole was the major problem. The fact that the parts which had to correspond were national systems in sovereign states meant that this had to be done through international agreements.

1.2 Theories of international organization

Traditionally much of the study of international organization has been carried out within the fields of politics and international relations, and dominated by what is normally referred to as realist or neo-realist approaches. The failure of more idealistic theories to explain the world and to secure a peaceful world order through the League of Nations gave rise to the realist approach to international relations.[6] Though developed and fragmented since its beginnings in the late 1930's and 1940's, this school of thought is based on three basic, underlying assumptions. First of all the state is the primary unit of analysis. Secondly, states are seen as rational and coherent actors, trying to satisfy their self-interest in rather the same way the economic man

---

[3] Used by for instance by Kaijser (1994)

[4] Kaijser, (1994), ch. 3

[5] Kaijser (1994), ch. 3

[6] Lieber (1991), pp. 10-11

does in neo-classical economic theory. This also implies that states have consistent and stable preferences, and try to maximise their utility by calculating the outcomes in terms of these preferences for all alternative policies. The third assumption postulates that the interaction of human beings, and in extension therefore also of states, is inherently conflictual. States struggle to maximise their welfare, and international organisations are simply arenas for this struggle, where power is measured in the form of military and economic power.[7]

In the field of telecommunications, international co-operation is seen as a struggle between states for power over strategic high technology, where various power blocks, such as North America, Europe and Japan try to set the world's telecommunication order to suit their own interests.[8]

One illustrative example of the neo-realist approach to the study of international telecommunications is provided by Arnold and Guy (1986), who view international competition for technological supremacy as following the same mechanism as arms races, where one country's advances provokes another country's response, which then leads to further development in the first country and so on.

The authors also stress the enormous economic resources needed for keeping up this 'technology race', where the leaders above all are the United States and Japan. This means that smaller and less developed countries, in order not to fall too far behind, have to develop specialised 'survival strategies' and/or manipulate the rules of the international game. The big players however, also have an interest in bending the rules for their own purposes, and, what is more, they have the economic and military muscle to do it successfully.[9] Writing about the EEC's IT-policies in the first half of the 1980's, they further regard the process of finding a Community-wide set of policies as one where the member states are in a more or less open struggle for defining the policy in a way that would benefit them most nationally.[10]

One of the problems with the neo-realist paradigm which gradually has led to the development of competing models of explanation is its tendency to isolate

---

[7]Lee (1996), pp. 17-19
[8]Lee (1996), pp. 19-22
[9]Arnold & Guy (1986), ch. 1
[10]Arnold & Guy (1986), ch. 5

10

international affairs from domestic policy. Such inability to account for domestic determinants of foreign policy is by more liberal theorists seen as "...an important barrier to serious, systematic investigation of such relationships."[11] In an overview of EU integration theory, focused on telecommunications and electricity regulations, Schmidt (1997) puts neo-realism at one end of a scale, representing a strong focus on the autonomous powers of states, and where international agreements are regarded as purely intergovernmental affairs. Another, more institutional, school of thought in international organisation theory is that of liberal and neo-liberal theories. The basis for these theories is that international co-operation exists because of overlapping interests between states. One step further along the scale towards supra-nationalism is the constructivist school, where international organisations are seen as pursuing interests of their own. In her comparison between the relative success of the EU Commission's harmonisation initiatives in telecommunications, and their relative failure in harmonising the electricity policy of the Union, Schmidt leans towards an explanation that focuses on the powers of a supranational actor (the Commission), but required some degree of homogeneity in the national industries.[12]

The functionalist studies in liberal and neo-liberal theory stress the importance of overlapping interests between states, and the functions of the policy area to be regulated, to explain international co-operation. Whether an activity shall be regulated regionally, nationally or internationally depends on its function. As telecommunications are growing ever more international in themselves, it is quite natural that the regulation of them also becomes international in scope.[13]

The process of international co-operation is seen as a process in which a number of factors, such as regimes (i.e. concrete procedures and rules of decision making) and conventions (i.e. informal institutions) shape the outcome. Within this context the role of private actors and interest groups are also appreciated, besides that of the states themselves. The private actors however are often seen as representing purely technical standpoints, as opposed to the political positions of the states.[14] Later

---

[11] Kahler (1997), p. 1
[12] Schmidt (1997)
[13] Lee (1996), pp. 23-24
[14] Lee (1996), pp. 24-26

11

writings on the International Telecommunications Union, ITU, in the liberalist school have abandoned this view of private actors. In trying to do away with the division between 'technical' and 'political' aspects of regulation they argue that there is no such thing as a politically neutral technical regulation. From this perspective the relative success of states or groups of states in influencing the agenda of the ITU are studied, and the issue of what resources such influence is based on.[15]

One problem with many of these theories is their treatment of states as having stable, unitary preferences, and the separation of national and international politics. In as far as domestic affairs are included in the analyses they seem to focus mainly on institutional arrangements of domestic politics and the function of executive bodies. There is reason however to study how domestic politics have a real influence on international affairs as well. In an influential article from 1988, Robert Putnam tries to do this by including political parties, interest groups etc.[16] He describes the relation between national and international politics as a two-level game:

> "At the national level, domestic groups pursue their interests by pressurising the government to adopt favorable policies, and politicians seek power by constructing coalitions among those groups. At the international level, national governments seek to maximize their own ability to satisfy domestic pressures, while minimizing the adverse consequences of foreign developments. Neither of the two games can be ignored by central decision-makers..."[17]

Drawing on these theories, this study assumes the importance of other factors than purely technical in international telephony co-operation. The "technical" positions of member states are thought to involve economic and political interest as well, which makes the study of regulatory regime interesting, even to others than those studying dissemination of technology. The setting of standards is supposed to involve economic and political considerations, and standards themselves are seen as a type of institutions in their own right. When setting a standard, one also directs future development of the technology concerned. In this way technological

---

[15]Savage (1989), pp 6-7, 10-11

[16] Putnam (1988), p. 432

[17] Putnam (1988), p. 434

development is path dependent, and in the long run questions that may seem technical have other economic and political implications as well.[18]

The role of the state differs from that in classic realist literature in that states are not seen as unambiguous entities, acting like individuals with clear-cut, consistent preferences. Other actors such as private enterprises, often with international links in themselves, interest groups etc., are seen as being important in the shaping of international regulations.[19]

These theoretical foundations of my research provide a basis for studying the interrelations between the national and international policy levels of telephone regulations. The view of standards and regulations as economic and political matters as well as technological; and the view of national positions being shaped by other considerations than rational and technical utility maximising, opens up the possibility of an institutional analysis of the regulating process.

1.3 Previous research on international telecommunications organizations

International telecommunications are a relatively little studied phenomenon of international relations. One reason for this might be that, with a few exceptions, issues of regulating the international telecommunications order have never led to high-profile international conflicts. The allocation of radio-spectrum frequencies to countries, where each country is allocated frequencies according to their political, social, economic and military needs is an exceptional case where the co-ordination of telecommunications suddenly becomes identified with power politics. In the post-war period the issue of harmful interference of radio and broadcasting is another high-profile example. But apart from that sort of exceptions, the international telecommunications regulation has seemingly been a case of co-operation rather than conflict.

George A. Codding Jr (1952) adheres to this view in his history of the International Telecommunication Union. In it he traces the development of the ITU from the formation of the Telegraph Union in 1865 up until the more mature and

---

[18]Savage (1989), pp. 12-15; Helgesson, Hultén and Puffert (1995)
[19]Lee (1996), p. 27

genuinely international body of the ITU in 1952. In his analysis of how the ITU has managed to continue functioning as an international body over such a long period, marked by international conflicts and two World Wars, he focuses precisely on the seemingly apolitical nature of its work. In Codding's view the key to the ITU's success lay in the fact that its work has been almost exclusively carried out by telecommunications experts. "Their disregard for legal formalities has at times led them into temporary difficulties, but at the same time it has prevented them from becoming so deeply involved in many of the ever present political machinations as to neglect their essential work."[20]

Codding further notes three other factors which contributed to the efficient operation of the Union, especially in the period before the Second World War. Firstly, he mentions the economic interdependence of Europe. The international character of the European economy and the density of population meant that there was a demand for international communications; and this demand in its turn meant that there was a real need for co-ordinating the national systems and thus enabling a coherent telecommunication network in Europe.

Secondly he regards as a facilitating factor that the telegraphs and telephones in Europe were mostly run by state agencies. In this way it became a matter of national pride, or a sign of economic and industrial maturity to be a member of the early Telegraph Union. Furthermore the telegraph and telephone Administrations' status as state agencies meant an advantage in implementing the rules when they had been agreed upon internationally. As support for this argument Codding notes that in the cases where, for instance the Atlantic telegraph, private companies operated, these often openly disregarded the rules relating to rates, laid down in the International Telegraph Regulations.

Codding's third factor deals with the organisation of the ITU. The structure of the ITU allowed it to decide on stable rules without failing to take account of a rapidly developing technology by dividing its work into two sections. Aspects of the international regulations that were of a diplomatic/political or of a more permanent nature were dealt with by plenipotentiary conferences. At these rather sparsely

---

[20] Codding (1952), p. 461

14

occurring conferences, delegates with plenipotentiary powers decided on the Telecommunication Conventions, intended to be of a long-term, stable character. More detailed technical or transitory matters were dealt with by technical experts from the member states' telecommunication Administrations. By in this way separating the technical from the political aspects of its work, and keeping the latter to a minimum, Codding concluded that the ITU had succeeded in its 'experiment in international co-operation'.

This view of the apolitical nature of international telecommunications co-operation as a key to its success can be found among other writers as well. One author writes about the shared project between the British Post Office and AT&T to design and build the first transatlantic telephone cable in 1953, that it "...provides an encouraging example of the way in which international conferences can succeed if only the politicians can be kept out of them"[21]

Kenneth D. Grimm (1972) diverges slightly from the view of the ITU as a non-political body, although he describes it as a, indeed the oldest existing, 'functional organisation'.[22] Functionalist studies indicate that international agencies should meet the criteria of being technical, functionally specific and essential. By this is meant that an agency is technical when there is a sophisticated technical knowledge required for its work; it is functionally specific when it relates to one specialised area of public policy; and essential when the performance of its tasks are best carried out internationally, i.e. when the states concerned have similar problems which they are trying to address.[23]

Grimm concludes that the establishment of the ITU was caused by overlapping technical interests of the European founding states, and that the reason for the organisation's survival has been due to the continued convergence of interests on behalf of its members. In his analysis of the ITU's work during its first one hundred years of existence, he qualifies the notion of the Union's work as being entirely technical and hence non-political, by indicating that international political

---

[21] Clarke (1958), p. 159
[22] Grimm (1972), p. 1
[23] Renaud (1990)

controversy occasionally is injected into the Union's affairs. One example of such political controversy came in connection with the allocation of radio spectrum frequencies to countries.

One key issue however, according to Grimm, is that the members themselves *perceive* the ITU's work to be unrelated to their political concerns. This, together with the more rationalist explanation that all participants are better off if they co-operate universally[24], works as an incentive for the members to keep their work outside the sphere of international political controversy.[25]

In another, more critical study of the history of the ITU, Kelley Lee (1996) tries to break away from the view of the Union as a non-political, technical body. In her analysis of the ITU's early history, she concludes that the founding principles of the ITU were wholly in accordance with the liberal ideas prevailing in 19th century international relations in Europe. Through the widespread use of the telegraph, the increased international flow of capital, goods, and services would enrich the participating countries. Voluntary membership and the principle of one state, one vote were features well in tune with the liberal ideals of the era.[26]

That was however soon to change by a more protectionist, nationalistic climate; and in the new situation the ITU took on the role of a colonialist vehicle. One example of this was the colonial voting rule. Introduced in 1875 on British initiative, the rule allowed member states to vote for territories under their rule. This meant that in 1925, Britain together with France, Italy and Portugal had seven votes each. The colonial voting system was not abandoned until 1973.[27]

Another example is provided by the shape of the international telegraph network before the Second World War, where the European colonial powers functioned as central hubs in relation to the more peripheral colonies, which in their turn had little communications directly with each other.

---

[24]That such a situation is not a satisfactory condition for actual co-operation is well known in e.g. game theory. 'The Prisoners' Dilemma' is just one example of situations where individual utility maximisation leads to sub-optimal outcomes collectively. See for instance Lieber (1991), pp. 263-267

[25]Grimm (1972), p. 3

[26] Lee (1996), ch.1

[27] Lee (1996), p. 60

Others have used the concept of corporate actors to discuss the case where actors, such as states, transfer some of their authority into a common body.[28] This can be done to create a stable basis for collective action, in order to achieve some common good. Once established however, these bodies start achieving their own preferences and aims, not always compatible with those of their original founders. They start living a life of their own, as it were. One striking example of this is the EU Commission bringing the EU Council to court for not implementing the common market for road transports.[29]

In the field of telecommunications Schneider, Dang-Nguyen, and Werle (1994) have applied the corporate actor concept to analyse the European telecommunications policy as a multi-level game where the EU institutions are seen as players in their own right, on a supra-national level. In their article, they analyse strategies by the EU Commission to harmonise the integration of European telecommunications further and faster than the responsible ministries in the member states were willing to go. They further refine their method of analysis by allowing for influences from a number of different interests, domestic and international, both on the national positions of member states, and directly on the supranational actors.[30]

The problem with many of these theorists is, as mentioned above, that they regard the international agreements on telecommunications as technical, and therefore non-political, matters. In some of the cases the authors introduce some measure of 'politics' in their analysis. This is however often done in an *ad hoc* way, by arguing that some of the issues in the work of the CCIF or ITU also have political implications.

In contrast to this, what I want to argue is that the technical issues in themselves are inherently political. Matters of economy, national security, culture etc. which are normally labelled as 'political' all have some influence over technical matters as well. This influence can take a number of different forms, such as legal restrictions to uses of technology, or tastes and demand. At the same time technological

---

[28]Coleman (1990)

[29]Bergdahl (1996)

[30] Schneider, Dang-Nguyen & Werle (1994)

development enables some paths of development while restricting others, in a way that in the long run affects most other areas of life too.

Technological development does not occur in isolation from other spheres of society. In the same way regulation of technology, or the 'politics of technology' is not isolated from other political issues. In order to understand it we must therefore treat these technical matters as part of the whole political system. Thus when dealing with national communications policy this must be related to other political concerns to make any sense. In the same way the study of international regulation of telephony requires the inclusion of other international relations.

By looking at the process of establishing a common set of rules for the international telephone system in Europe as a game, I want to investigate which forces affected the British position, and how Britain, as a player in this game, acted to overcome or make use of those forces. In other words I want to answer the following questions: if Britain acted strategically within the CCIF, what strategies did she use?; and what determined that particular choice of strategy?

As I claim that telephone policy in fact is no different from, but interlinked with, other political areas, I endeavour to show how that interrelatedness was manifested. In terms of my work in this thesis, that means studying which other fields of politics were influential in shaping the British telephone policy. This also involves studying which actors were involved in the political game. Thus two additional questions can be formulated: What constituted the 'British position' or British interests; and how was that position reached?

Apart from these theoretical issues, another reason for my choosing to study this particular question is the lack of research carried out in that empirical field. Although a number of studies have analysed the international co-operation within the CCIF and ITU, none have specifically studied the interplay between national policies and telecommunication co-ordination. Neither have I found any coherent study dealing specifically with Britain's relations to those international bodies. There is therefore a lack of knowledge in these fields, which I will try to rectify with this thesis.

1.4 Aims and scope of this thesis

The overriding aim of this thesis is to study how one member state, Great Britain, acted to put forward her national interests in order to influence the whole of the European regulatory system; and how those national interests were formed on the national level.

As for the limits of the work presented here, I have chosen to study the strategic behaviour of one country only. This has to do with the embeddedness and interrelatedness of telephone policy to other political fields. One of the points to make is that the analysis of one political arena can not be made in isolation from all others. The in-depth analysis of domestic as well as international actors and political forces that this requires, necessitates that it is carried out as a case-study, and the limits set for a licentiate thesis means that it is realistic to select only one country. The reasons for choosing Great Britain include that Britain was an important player in European telephony as well as world politics who in many ways was and perceived herself to be, different from continental Europe.

The chosen period of study is 1923-39; that is, from the formation of the CCIF (or its immediate predecessor) up until the outbreak of the Second World War. As usual, these limits are not absolute. In many cases the development before 1923 has to be included in order to explain why things turned out as they did.

The reason for stopping at 1939 is the dramatic effect the War had on the European telephone system. During the hostilities, much of the telephone network in Europe was destroyed, international telephony was limited or non-existent within Europe, and the CCIF did not conduct any work. After the War, the new European telephone system was rebuilt from new principles. The CCIF became incorporated with the United Nations, and much more of an international rather than European body. At the same time the influence of the United States replaced the dominant position earlier held by the European states within the organization.[31]

One important point to make regarding the limits of this work, is that this thesis is intended to be one part of a larger study. In this part I will study how Britain acted to influence the CCIF, without paying much attention to how the CCIF in itself acted

---

[31] Chapuis (1976), pp. 192-3

strategically. With the notion of the CCIF as a corporate actor with preferences of its own follows the need to study how those preferences were formed, and which strategies the CCIF used to forward its interests. This other part of the equation will be the subject of my coming doctorate thesis.

One last word of the limited scope of this study. At the risk of being repetitive I want to reiterate that the aim of the thesis is to study the strategic behaviour of Britain. This means that the results of this behaviour are not the primary focus of interest. In order to estimate how successful Britain was in her efforts, one would have to weigh the actual result of the co-operation against some counterfactual outcome in the case that Britain had acted differently or not at all, and that goes outside the ambition of this work.

1.5 Defining the actors

The term 'actor' in social sciences normally signifies a decision-making unit of some form, e.g. individuals, organizations, or states.[32] From the standpoint of methodologic individualism it can be problematic to regard complex units such as organizations or states as coherent actors with unitary preferences.[33] Indeed one of the points I want to make in this study is the necessity to analytically break down the notion of national positions into a number of different interests which together shape what is then known as a national stance.

Yet this still does not meet the strictest demands of methodologic individualism. There are however two reasons for stopping the analysis from the organizational to the individual level. The first is simplification. The complexity of the analysis would increase immensely at the expense of clarity if all the actions of organizations and ministries had to be broken down to individual motives. The second reason is related to source material. The archive material I have studied contains reports, memos, and letters written by people in their professional capacities. Trying to establish whether opinions expressed in these were the result of personal preferences or the 'official' position of the organization they represented would require different sources. For

---

[32] Goldmann, Pedersen & Østerud (1997), p. 11
[33] Hovi (1992), pp. 18-22

20

these reasons I will, with a few exceptions, treat bodies such as the Treasury as having unitary preferences.

The use of the metaphor of a game for describing political processes is rather common among political scientists. In this thesis there will be no use of formalised game theory however. The purpose of the game metaphor here is only to illustrate and highlight a number of key features which politics share with more normal games.

One such feature is the fact that there normally are a set of rules that must be followed. In some cases these rules can be changed during the game, but that typically includes convincing others that the changes are for the better for them too, and can thus be included as part of the game as well. There is also a number of different actors involved, who more often than not have diverging goals. Most importantly however, the individual outcome for any of the players is the result of their own actions *as well as* those of the others. The overall outcome of the game is decided by how well the different actors use their relative strengths and take advantage of the opportunities that arise for achieving their goals.[34]

Before venturing into the details of the process of international telephone co-ordination it can be useful to define who were the players in this game.

The Post Office was the body entrusted with running the state owned national telephone system after it was nationalised in 1912. In relation to the formation of the British objectives we can not however regard the Post Office as an unambiguous actor. Within the Post Office there were what can be described as two factions; one 'expansionist' party in favour of extending and improving the telephone system as a way of improving the national communications system, and one group of 'restrictionists', unwilling to invest in what they regarded as a potentially risky system with doubtful advantages. In the early part of our period the Post Office, apart from its initially small telephone branch, could be described as restrictionist but gradually things shift towards expansionism.

Within the British Government there were other forces not principally involved with communications politics who nevertheless had some influence over the

---

[34] See for instance Hovi (1992), ch. 2. For a slightly more popular but immensely useful treatment of politics as games, see Laver (1997).

formation of British objectives. The Treasury was naturally concerned since any investment in the telephone system required state expenditure. The Treasury was directly involved in any major investment decision of the Post Office and accounted for the net financial result of its operations.

In as far as the telephone co-operation involved making agreements with foreign states, this was also the concern of the Foreign Office. During the period the Post Office gradually gained the power to negotiate agreements directly with foreign telephone administrations but they still submitted drafts of these agreements to the Foreign Office to decide whether they should be executed through diplomatic channels. The Foreign Office could also give general directives to British delegates, for instance that they should keep in close touch with the Dominion delegations.[35] On a more general level the official British relations towards other states were also affected by the judgements of the Armed Forces.

Most of Britain's representatives in the CCIF were engineers and members of the Institution of Electrical Engineers, an organization for discussing engineering matters. Many of the influential members in that organization were American telephone engineers from the Bell System and AT&T. Through their personal networks with the British telephone engineers in the Post Office and the high regard these held for the American telephone system, the Bell engineers had some influence in shaping the British objectives as well.

Within the CCIF, the other players principally were the representatives of the other member states, who in their turn had similar networks of different interests shaping their national positions. One feature which quickly becomes apparent when studying the CCIF negotiations is that whenever there is a conflict with countries disagreeing over something, Britain seems to be on one side and Germany on the other. In most cases however, the formation of blocs appear to vary with different issues.

---

[35] Instructions to the United Kingdom delegation to the Telegraph and Radiocommunication Conferences of Cairo, 1938, pp. 11-14

In as far as the CCIF can be seen as a player in its own right, a corporate actor, this will be the subject of the coming doctoral thesis and not specifically dealt with here.

1.6 Methodology and sources

Methodologically the study of British strategic action within the CCIF is a process in two steps. Firstly, I want to isolate such a 'British position'. There is of course no document outlining that position. Instead I will have to reason my way to it from the various factors which had direct effects on the international telephone policy of Britain, and study the forces which put pressures of various kinds on it. This is done by drawing conclusions both from British national communications policy, and on a larger scale British international relations, so as to define what Britain wanted to achieve regarding the international telephone network. The process of reaching such a national position was often a highly complex matter, where different actors on the national arena held differing views on what should constitute the 'national interest'.

In Britain, as in most other European countries, the telephone system once it was in the hands of the state, was operated by the Post Office.[36] In my analysis of the formation of British objectives, my main sources are correspondence between the Post Office and other bodies, internal memoranda, and internal reports. The analysis of that material is greatly helped by the good shape of the archive from the telephone branch of the Post Office, now held at BT Archives in London. The archive in most cases contains a number of draft versions and background material for the reports and memoranda, as well as 'both sides' of the correspondence, all filed after their subject matter. My overview of the organisational history of the Post Office also draws heavily on Pitt (1980) whose treatment of the subject, after a number of random cross-references, seems to follow his sources closely.

Secondly I turn to the actual European co-ordination and co-operation to study what kind of strategies Britain used for accomplishing those ends. Official protocols from the meetings of the CCIF are held both at the Public Records Office, and in Televerket's (the Swedish national operator, now Telia) archives, now held at

Landsarkivet in Uppsala. Internal Post Office reports and correspondence give 'the British side' of the CCIF negotiations. The BT and Televerket archives also contain reports from the various expert study groups, which tend to be more open and refer more of the internal discussions than the protocols from the Plenary Assemblies of the CCIF.

1.7 Structure of the thesis

After having dealt briefly with my aims and methodology of the thesis in this introductory chapter, and with the theoretical background against which the analysis is set, Chapter 2 will go on to discuss how international telephony in Europe was regulated in the early 20th century. In that context the early technology of long-distance telephony will be discussed, to serve as an illustration of what sort of limitations technology gave to the relatively new means of communication. Chapter 3 deals with the development of an institutional structure for regulating the international network; how the CCIF evolved and what actors were behind it.

Turning then more specifically to the case of Great Britain, Chapter 4 tries to isolate some British 'objectives': persistent features in the way Britain acted in the international co-operation and what the British representatives tried to achieve. These objectives are identified by studying various aspects of national and international policy questions that are wider than the specific field of international telephony, but still influenced British preferences. In Chapter 5 I show which different strategies Britain used within the international co-operation to put forward her objectives, and how different objectives required different strategies. A concluding discussion is found in Chapter 6.

---

[36] The term normally used for that kind of public communications operator is PTT, short for Posts, Telegraphs, and Telephones.

2. The development of the European telephone system in the late 19th and early 20th century

The object of this chapter is to describe the institutional set-up of the European telephone network prior to the formation of the CCIF. This will give a background against which the later co-operation and co-ordination efforts can be measured, and serve as an aid to understanding the nature of the co-ordination problems that had to be solved. By 'institutional set-up' I primarily mean two factors. Firstly the hard, technological side of long-distance telephony during this period, and secondly the political or organisational system through which the process of co-ordinating the national systems took place.

The two are interlinked in as far as the processes of technological development had given rise to national differences in telephone technology, and thus so to speak had created the need for co-ordination. Both factors also, in their own way, set the limits to the expansion of an all-European telephone network. As we shall see, technological developments gradually enabled telephony over longer distances and thus both created the opportunities and set the limits for a coherent European network. Technological development was however a necessary, but not a sufficient condition for such a system. In order to realise the possibilities offered by technology, the European states had to organise some form of co-operation between their telephone operators.

2.1 Technological development

As has been mentioned earlier, the technology of long-distance telephony had developed rapidly in the first decades of the 20th century. The technological development was the factor that made telephony over long distances possible in the first place, and to some extent set the 'hardware' limits for the expansion of the telephone network. It can therefore be useful to study, at least briefly, the technological developments in the early 20th century, so as to describe the framework against which the political and organisational development was set.

Technological development of long-distance telephony naturally had to precede economic demand for it. At the same time however, technological development, just like social or political changes, is not something that occurs by itself. Frank Knight describes the interaction of demand and other economic factors thus:

> "Wants are usually treated as the fundamental data, the ultimate driving force in economic activity, and in a short-run view of problems this is scientifically legitimate. But in the long-run it is just as clear that wants are dependent variables, that they are largely caused and formed by economic activity."[1]

And technology, one might add. Technological development does not proceed automatically according to some predetermined rational masterplan, but neither are the various small steps that together constitute technological development taken randomly or haphazardly. Thomas Hughes describes the process as one where the whole technological system holds reverse salients, or imbalances, which direct human action towards redressing these imbalances through invention, reorganisation etc.[2]

These innovations in their turn give rise to new reverse salients, which will direct future inventive efforts. Thus the development is not a mechanic process, but depends on what is perceived as imbalances. In this way the technological systems become embedded in social life. Technological development is in a way formed by social norms and human behaviour. In the same way does technology shape future social norms and human behaviour.

The embeddedness further means that technological development is path dependent. The path dependency facilitates future development by directing efforts in roughly the same direction. At the same time however, it restricts the possibilities of choice for that future development. "...[a]s the past technologies hold a stronger grip over the present than all future possibilities, old technologies will certainly affect the development of new technologies replacing them."[3] This gives rise to technological styles, where for instance legislation, geographic factors or historical circumstances all contribute to different solutions to technical problems, and hence to alternative technical systems in different locations.[4]

---

[1]Knight, F., (1924), pp. 262-3, quoted in Hodgson (1988), p. 20
[2]Hughes, (1987)
[3]Helgesson (1994), p. 29
[4]Hughes (1987)

26

One illustration of this is the situation in European telephony during the early 20th century, where interconnection between the various national networks was highly limited, in part due to the differences in technology and practices.

2.1.1 Telephone technology

The basic technology for telephones can however be described as a common ground for later developments. When Alexander Graham Bell filed his famous and debated patent application for the first telephone, on February 14, 1876, its basic principles were the same as would be used for most of this century, and to some extent are still used.

Schematically a telephone call can be divided into a number of different stages on its way from one subscriber to the other (see figure 2.1)

Figure 2.1 Separate functions in a telephone call

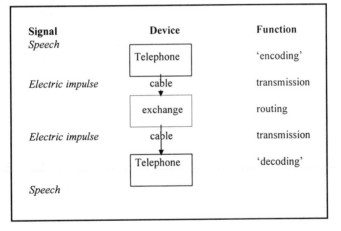

The telephone has a microphone with a diaphragm or other physical device, which is set in motion with the same amplitude and frequency as the sound waves of speech directed into it. An electric current is made to covary with the oscillations of the diaphragm, thus 'encoding' the sound waves of speech into an electric current. The electric impulse is then transmitted over a telephone line and possibly a telephone network to the receiving set. There the diaphragm of the receiving telephone's loudspeaker again covaries with the current to 'decode' it, i.e. produce sound waves so that the speech from the sender can be heard.

2.1.1.1   Telephones

Over the decades following the invention of the telephone, various improvements were made to the original design so that gradually the quality of sound improved and the maximum range of a telephone call was extended. One such fields where important improvements were made was that of the microphones in the telephone sets. Microphones which delivered up to ten times stronger electrical currents than the normal ones were tried and found successful. The principal advantage with a stronger microphone was that it could produce a much more powerful signal, and thus lessen the problem with subdued electrical currents.

By also improving the signal, i.e. lessen the level of noise in relation to the speech that was transmitted, the audibility at the receiving end could be improved. The original microphones were rather insensitive instruments, with dramatic variations in the levels of sound. A refined version of the microphone, which remained in use long into this century, introduced a layer of carbon granules between the diaphragm and the magnet. In that way a more modulated sound with higher precision in reproducing the original sound was achieved.

2.1.1.2   Cables

Another part of the telephone system which had a great impact on its use was the development of transmission systems. The early telephone networks were local affairs. The networks used a single-cable system where the electric impulse was strongly subdued, giving a maximum range for phone calls of some 100 kilometres. Towards the end of the 1880's a new double-line technology was developed, which made conversations up to 1000 kilometres possible.[5] Still transmission over long distances was a great problem since the electric current was heavily subdued, and the quality of sound poor.

In 1899 the electrical engineer Michael Pupin discovered that the problem with subdued currents could be reduced by inserting coils of copper wire with regular

---

[5] Kaijser (1995)

28

intervals of 1500 metres on the telephone lines. These 'pupin-coils' revolutionised long-distance telephony, both by enabling telephony over greater distances than before, and by making long-distance cables much cheaper to construct.[6]

The problems associated with the fact that the signals were strongly subdued over long distances was still present though. The 'pupinisation' of telephone cables had lessened the problems, but not solved them. The solution came from radio technology and electronic valves. The first amplifier was patented in 1906, and primarily intended for radio receivers. Its inventor, the American Lee de Forest, seems to have appreciated that it could also be used for telephony, but the first functional telephone repeater was not constructed until 1910, by the Austrian engineer Robert von Lieben.[7]

The principle behind the telephonic repeater was that variations in the incoming electric signal caused analogous variations in the currents from an internal anode battery, which were considerably stronger. Thus the received telephone signal could be repeated down the telephone line with a much higher amplitude. By connecting three repeaters in a series, de Forest showed in 1912 that the amplitude could be increased 27 times.[8] Theoretically, the telephone repeater would mean that geographical distance would lose its importance for establishing telephone contact. In the United States, AT&T put the repeater to the test when they in 1915 established a telephone line between New York and San Francisco. By inserting a sufficient number of repeaters at the right intervals a telephonic message could, in theory, be transmitted around the world. In practice however, things were not so easy. One of the main problems was to insert repeaters in submarine cables. It was both difficult and expensive to construct submarine cables with good quality repeating, especially over long distances, and until 1956 all intercontinental telephone traffic had to be transmitted via radio.

---

[6]Heimbürger (1953), pp. 27-30
[7]Heimbürger (1953), p. 81
[8]Heimbürger (1953), pp. 81-2

2.1.1.3 Exchanges

In order for a telephone call to reach its destination, it must be routed from the calling party to the receiver. The earliest telephone systems were simply made up of two telephones connected with a cable. As the networks grew more complex, the need for more advanced exchanges arose. Originally the exchanges were manually operated. The caller alerted the operator by sending a signal to the exchange station, who could then connect the call to the wanted receiver; and when the conversation was finished, the caller again indicated this to the operator who could disconnect the lines.[9]

This system needed a battery or other source of electricity at each telephone, and was also dependent on the callers themselves to report when the conversation was over. An improvement was the introduction of the central battery (CB) exchanges, where the electricity needed for signalling between the caller and the operator was supplied from a central station, and where the operator automatically could see whether a conversation was finished.

As the use of the telephone became more widespread , the size and complexity of the exchanges increased. As the telephone networks in the larger cities developed, a complex set of hierarchies of exchanges had to be constructed. The manually switched telephone networks were often troubled by lack of capacity at peak times, and errors in either connecting the calls or of terminating them at the wrong time. One of the first automatic exchanges, created by a Mr A. B. Strowger in 1889, is said to have been the result of its inventor's sheer frustration over mistakes at his local exchange.[10]

The development of automatic exchanges was one field of technology where national strategies diverged. The technology based on the step-by-step principle developed by Strowger was initiated in the United States, who thus took the lead. The Bell system, though they developed the technology, chose not to install automatic exchanges on any greater scale until after the First World War, and in Britain the Post Office followed the lead from Bell, though with a time-lag of a

---

[9] Jacobaeus (1976), pp. 58-67

[10] Larsen, (1977), p. 38; Larsen refers to it as the first automatic exchange. Already in 1883 however, a Mr Dane Sinclair, later Engineer in Chief to the National Telephone Company constructed the first automatic switchboard to be used in Britain. (Foreman-Peck, 1992, p. 168)

couple of years. Instead German engineers strongly believed in the possibilities of automatic switching, and were early both in developing systems of their own and in installing them.[11] The rest of continental Europe lagged behind Germany. Sweden for instance, where LM Ericsson had developed their first semi-automatic exchange in 1883, did not introduce automatic exchanges to any greater extent before the 1920's.[12]

A number of factors could account for the different attitudes towards automatic switching. However, these do not quite satisfactory explain why the development and use of the new technology was most progressive in Germany. First of all, compared to manual exchanges, the automatic technology is intensive in capital costs but saves on operating and maintenance costs. That suggests that USA, with its higher labour costs, should have a stronger incentive to introduce automatic exchanges. Foreman-Peck (1992) suggests that "… a lesser German emphasis on short term commercial results and/or a higher target quality of service accounted for the difference between the systems"[13]. Secondly, calls for other exchanges still had to be switched manually. Hence the benefits from introducing automatic exchanges depended on the proportion of local calls, i.e. calls which originated and terminated within the same exchange area.

2.1.2 National differences in technology

As we have seen, telephone technology started as a relatively homogenous technology. Gradually however, national differences started developing as a number of countries tried to develop their domestic telephone industry. This was perhaps most marked in Germany and Scandinavia, whereas the Bell system controlled much of the industry in the rest of Europe. In Britain the telephone companies also looked closely at Bell, which led to similar technical solutions, albeit with a time-lag of a couple of years. In that way it could be argued that the Bell system acted as a unifier in European telephone technology as well.

---

[11] Foreman-Peck (1992), pp. 168-9

[12] Jacobaeus (1976), pp. 77-83

[13] Foreman-Peck (1992), p. 129

The main differences in technology are found in the exchanges. Not only did the European countries differ in timing when they introduced automatic switching. In many cases, but again above all in Germany and Scandinavia, they chose to develop their own systems. Arguments for this seems to have concerned both industrial policy and nationalism. In cases where the state had a positive view of the possibilities of the telephone of becoming a major new industry, clearly there was some economic incentives for trying to develop a technological leadership in that field. That also went hand in hand with the view of telephony as a feature of 'modern' society, and hence the desire to establish a domestic telephone industry to demonstrate the modernity of the state.[14]

For most of the European countries, telephone network construction followed a roughly similar phase, though with some variation in timing. The construction of local or urban networks began in the late 1870's and gathered pace in the 1880's. In the two decades around the turn of the century those networks were gradually connected into national systems. There the interconnection halted however. Some single lines between the national networks were established from around 1900, but until the mid-1920's, when the CCIF began its work, there is no justification for talking about an international network in Europe.

This lack of interconnection can not altogether be explained by heterogeneity of technology. To be useful, the concept of heterogeneous technologies must be qualified. There is no clear-cut dichotomy between homogenous technologies, which can easily be interconnected regardless of geography, and heterogeneous technologies which simply can not be interconnected. In almost any case the question of interconnection of two systems requires some amount of reorganisation or technological adjustment of them. In that way the homogeneity of a certain technology could be expressed as the cost of such adjustments. We can thus think of it as a scale running from homogenous technologies (i.e. requiring small adjustment costs), to heterogeneous (i.e. requiring larger adjustment costs to be interconnected). The question is then where to place long-distance telephony on that scale.

---

[14] Helgesson (1995)

In the case of telephones there surely was some amount of heterogeneity in the system, due to the national styles that had evolved. However, for the purpose of establishing an international network there was no direct need to conform all parts of the national systems. In most cases the international or long-distance lines ran separate from the local lines, and thus the need was primarily for enabling switching between the long-distance and the local networks. Thus, for the initial programme of interconnection of the European system the adjustment costs in relation to the benefits from it, were relatively low.

2.2 International agreements

In the early years of telephony, there was no international body to deal specifically with the issues arising from telephony. The international organisation perhaps closest to hand to deal with the international co-ordination of telephony, the International Telegraph Union (ITU), showed relatively little interest in doing that. At the international telegraph conference in Berlin in 1885 some regulations regarding telephony were added to the international telegraph convention. These were however of a rather general nature. They stated that telephone traffic between countries could take place if the concerned states had so agreed, that the conversations should be charged per 5 minute period, that the conversations should be handled in the same order that they were requested, and that no conversation longer than two periods should be permitted if other requested calls were waiting.

These general regulations were supplemented with more detailed rules at the international telegraph conference in London in 1903. For example the new regulations included the introduction of zone charges, i.e. the practice of dividing each country into a number of geographical zones, and then determining the total charge by adding the zone charges. This gave a more flexible pricing of telephone conversations, which also more closely converged with the costs of handling a telephone call. Other regulations that were introduced were lower rates for night-time traffic; and special traffic categories such as express calls at higher rates, and calls of State with precedence over all other traffic. At the telegraph conference in Lisbon, the telephone regulations were further added with instructions for 'avis d'appel', i.e.

the service that a certain person could be summoned in beforehand to a certain telephone for receiving a telephone call.[15]

These regulations were of a very general character, and in fact did little for the actual establishment of an international network. The international telephone lines that existed prior to the 1920's were in most cases based on bilateral agreements. The practices of negotiating and establishing agreements with foreign countries varied between the different European states, and the absence of all-European rules as to what should be included in the agreements led to rather complex sets of rules for international telephony, where certain services were available to some countries and not to others etc.

2.2.1 Early Anglo-Continental lines

For Britain, the first discussion of establishing telephone contact with the continent dates to 1889. In that year the desirability of establishing such contact between Paris and London was expressed by the French Minister for Posts and Telegraphs, and the matter was immediately taken up by the British Postal Authorities. The chief problem was to establish a telephone line across the Channel. Initial attempts to use the already existing telegraph lines for telephony failed, which led the British Post Office to order a submarine cable from Messrs. Siemens Bros. and Company. Through this cable, the first telephone contact between Britain and the continent was opened, with a line between London and Paris, on April 1st, 1891. This, then, must be said to be the beginning of international telephony in Britain.[16]

One problem of the service, which was to remain a problem for a long time in European long-distance traffic, was the persistent feature of overcrowding of the telephone lines, due in part to the low price policy of the European telephone administrations. This was true also of the Anglo-French line, so that the need soon arose for extension of the service. In 1897 therefore, two new cables were laid across the channel; one by the British Post Office and the other by the French

---

[15]Heimbürger (1953), p. 40
[16]Baldwin (1925), pp. 481-2

34

administration. This extended the Anglo-French telephone service to include also the principal towns of England and the north of France.[17]

A further development of the British international telephone contacts was made in 1902, when a cable between Britain and Belgium established telephonic contact between London and Brussels.[18] An agreement between the Swiss, French and British administrations in 1913 led to the opening of a service to Switzerland, switched at Paris. The London-Paris line however, was almost constantly congested and therefore slow and unreliable as to whether calls could actually be completed, meaning that the Anglo-Swiss service in practice was of very limited use until a direct "through" circuit between Britain and Zurich and Basle was established in 1927.

One important fact to consider, regarding British international telephony, was that, though early in its inception, it effectively only allowed communication between London and Paris, the north of France, Brussels and Antwerp by 1922. It is against this background the frustration of telephone officials and the call for international co-operation must be seen.

The development in the 1920's of telephonic repeaters and thermionic valves dramatically reduced the cost of trunk lines. This, in connection with the development of a co-operative structure between the European telephone administrations, meant that the European telephone systems during the course of the decade came to be connected into what was called the European system.[19]

2.2.2 British international agreements

As mentioned earlier, the telephone operator in Britain after the state had taken over the national system was the Post Office. In Britain this can to some extent be explained by institutional path dependency. When the state decided to nationalise the

---

[17] Post Office (1897); Post Office (1904)

[18] Baldwin (1925), pp. 496-7, 502-3

[19] As defined by the International Telecommunications Convention, the European system "comprises all the countries of Europe, and countries outside Europe the administrations of which declare that they belong to this system." , International Telecommunication Convention, Revision of Cairo, 1938, Ch. I, art. 2. This was an administrative definition of the system, in order to incorporate other telephone administrations into a co-operation that originally was a limited organisation of western European countries.

telegraph system in the 1860's, the monopoly power to operate national telegraphs was given to the Postmaster General. When the turn had come to the telephone system to be nationalised, the strategy used by the state was to claim that a telephone really was a form of telegraph, and therefore fell under the Post Office's monopoly rights.[20]

In most European countries the state operated the telephone system through such PTTs (Post, Telegraph, and Telephones). Sweden was one of the exceptions from the rule, where the telephone and telegraph systems were operated by a separate body, Telegrafverket. The issue of merging Telegrafverket into the Swedish postal operator, Postverket, was however on the Swedish political agenda in the late 1920's but came to nothing partly due to fierce resistance from Telegrafverket who claimed that operation of posts and telecommunications had absolutely nothing in common.[21] The whole issue of the political implications and political goals expressed in various agreements on telephony with foreign powers, naturally depends on what extent of discretionary power was given to the Post Office. Given the assumption that the Post Office represents the 'technical' point of view, as opposed to the 'political' one of Parliament and Treasury, it is important to sort out in which cases the Post Office could enter contracts directly with their foreign counterparts, and when the approval of the political sphere was necessary. Important to note in this context though, is the fact that the Post Office was a State Department, and that for most of the time concerned, the Postmaster-General was a member of Cabinet.

The primary reason why Parliament should have a say in the establishment of contracts regarding telephone and telegraph matters was of course that the said agreements would be made in the name of Government, and that any costs emanating from them would have to be met from public expenditure. In the early years of international telegraphy, the House of Commons concluded that all contracts "...extending over a period of years and creating a public charge...entered into by the

---

[20] For a further discussion on this, see ch. 4.2.1 Legal Post Office monopoly.
[21] Heimbürger (1968), pp. 20-21

Government...for the purpose of telegraphic communications beyond sea..." would have to be approved by a Resolution of the House in order to be binding.[22]

This however, seems to have applied only to contracts with individuals or corporations, which could give rise to questions before British Courts of Justice. Treaties with foreign governments were understood to be concluded on the responsibility of the Government, and could not be enforced in British courts. Therefore the opinion of Parliament could only be expressed after the treaties were already in existence.[23]

In 1889 the Foreign Office and the Post Office agreed that all international engagements regarding purely telegraph matters should be negotiated by the Post Office and signed by the Postmaster-General. In this context there also developed a difference between the terms 'Agreement' and 'Convention'. The purely telegraphic matters taken care of by the Post Office should be called Agreements, whereas the term Convention should be reserved for those matters being negotiated through diplomatic channels. Later the use of the term 'Convention' was restricted further, to include only those agreements which were concluded between the Heads of States, meaning that also those matters going through the Foreign Office and being concluded between Governments, were to be styled Agreements.[24]

It was still the practice in international matters though, to submit a draft of all international agreements negotiated by the Post Office to the Foreign Office to decide whether or not they should be executed though the diplomatic channels. As regards the influence of the Treasury, the rule was that the Foreign Office always submitted draft agreements to the Treasury. In the case of the Foreign Telegraph Branch, they normally submitted drafts of international agreements to Treasury when they involved matters relating to public expenditure. Since this has normally been the case in telephone matters, both the Treasury and Foreign Office having been

---

[22]Extract from Notes and Proceedings of the House of Commons for Wednesday 14 July 1869, quoted in Post Office (1887)

[23]Ibid.

[24]Post Office (1906) In the case of the opening of telephone services between Britain and Belgium in 1902, the agreement reached is called 'Telephone Convention'. This despite the fact that it was negotiated by the Post Office and the Belgian Telephone administration, and signed by the respective Ministers for Foreign Affairs.

involved in the process of negotiating agreements with foreign administrations, before they have been signed.[25]

The role of Parliament was in such way reduced to being informed of the agreements made, after they had been signed. Initially this was done by the Secretary to the Treasury, but when matters regarded purely Post Office business, it was decided that the Postmaster-General, when being a Cabinet minister with a seat in the House of Commons, would himself present these matters to the House.[26]

The first agreement regarding international telephony; the Anglo-French telephone service agreement in 1891, was negotiated and signed by the Post Office and the French telegraph administration. The draft agreement was, however, submitted to Treasury for acceptance of the agreed rates, and to the Foreign Office for acceptance of the whole agreement before it could be signed. At the request of the French government a declaration, confirming the agreement, then had to be executed through the diplomatic channel and signed by the respective Ministers of Foreign Affairs.[27]

2.3 Summary

In this chapter we have touched the subject of technological development in general, and the development of technological national styles. Even in the case of such a rather new technology as long-distance telephony in the first decades of the 20th century, different nations had made certain choices in their telephone system which made them different from those of their neighbours. We have also seen that the technological advances of telephony gradually made it possible to make telephone calls over greater distances. In other words, the technological or hard-ware limits to what was possible in terms of a European network gradually lost their importance.

Instead there was a need for some form of co-ordination of the national networks. However in the absence of some international body to set common standards or practices for all European countries, in most cases they had to reach

---

[25]Post Office (1906)
[26]Post Office (1905)
[27]Post Office (1906)

bilateral agreements. Since these agreements also differed, depending on which countries were involved, the result was a heterogeneous and patchy system with limited interconnection. This ultimately led to the insight that in order to create a unified system in Europe, one would also have to create a unified set of institutions. In the next chapter we shall turn our attention to how this was attempted in the early years of the 1920's.

3. The setting up of the CCIF

As seen in the previous chapter, the European system in the early 1920's was far from a single, unified network. This chapter deals with the first attempts at approaching that co-ordination problem in a systematic and permanent way by the setting up of a common European organization for the regulation of international telephony, CCIF. Outlining the creation and early development of the CCIF will serve as setting the stage for the political game of telephone regulation. By answering questions of which issues lay within the competence of the CCIF co-operation, who its members were, and how its procedures of decision making worked, the aim is to give some impression of what kind of organization the CCIF was, and to what extent there was an opportunity for strategic action on behalf of the member states.

In chapter 2 we saw that the technological advances that had been made in the telephone industry during the first decades of the 20th century had made telephony over very long distances possible. In the United States a telephone line between New York and San Francisco, a distance of some 4500 kilometres, had been established in 1915; and in the beginning of the 1920's it was possible to make phone calls between all the parts of the country.[1]

In Europe the situation was wholly different. Sweden, being a relatively advanced telephone nation by international standards, had in the beginning of the 1920's telephone contact only with Denmark, Norway, Germany and the most northern parts of Finland. The American example clearly showed that it was not just technology that limited the opportunities for long-distance traffic. The problem was rather lack of capital for building a cable network for long distance telephony and maybe above all: lack of co-ordination between the countries still trying to get back to normal after the war.[2]

---

[1] Heimbürger, H. (1974), pp. 310-312
[2] Ibid.

One of the earliest and most influential callers for some form of co-ordination was Mr Frank Gill, chief engineer at the International Western Electric Company.[3] In his Presidential address before the Institution of Electrical Engineers in 1922 he pointed out that in Europe there were:

- "About 40 self-contained local operating organisations, each, in the majority of cases, conducting a local business and a through business within its area, also that part of the international through business which lies within its own borders.
- No organisation controlling or co-ordinating the various local operating organisations, which yet have to function as a whole.
- No means of keeping the separate organisations in touch with each other, and no systematic means of adjusting differences in matters of daily practice.
- No organisation of any kind which handles and cares for the through business as a whole.
- No common agreement as to manufacture
- No common research, standard practice or technique of construction, maintenance and operation."[4]

Having identified the problem, Mr Gill then followed to suggest three alternative solutions to it. His first suggestion was to establish a European long-lines company to operate all the international telephone traffic within Europe, under licences from the various governments. An alternate solution was to form a Commission, of which only governments could be stockholders, to carry the international telephony. His third alternative, and this was to regard as a temporary solution only, was for the different operating telephone authorities to form an association which would study the problems and needs of European long distance telephony; and make recommendations regarding technical standards, measurements, operating instructions etc., enforceable to the participating authorities.[5]

Seeing the lack of co-ordination and a coherent strategy in the European international network as its main problem, Gill naturally focused on the issue of how to resolve that. Following from that basic idea, his primary solution was to introduce a monopoly, either private or state-owned. In the absence of such a monopoly, the

---

[3] Already in June 1921 an article entitled "Long Distance Telephony in Europe" by M. G. Martin appeared in the "Annales des Postes, Télégraphes et Téléphones". It does not seem to have reached much attention though.

[4] Gill, F. (1924)

national operators would, as a second best result, start co-operating. Such an arrangement might well be described as a form of cartel-agreement, where the whole of the European market was subdivided into regional monopolies. Logically, these alternatives can be illustrated in the form of a four-field matrix, as in figure 3.1.

Table 3.1 Gill's alternative solutions for the European market

	Monopoly	Cartel-agreement
Public	Second suggestion	Third suggestion
Private	First suggestion	

In the following we will turn to how the CCIF was established and developed as an organization. The structure of this chapter is chronological, but three general themes can be distinguished. After the establishment of the CCIF followed a period of intensive co-ordination of technical matters between the member states. The initial structure of the organization was however rather loosely defined, so after a couple of tears the CCIF had to turn its attention to its own internal structure and relations to other bodies. Once this was done, gradually new issues of telephony were brought onto the CCIF's agenda.

3.1.1 Paris 1923

Following Gill's suggestions France invited its neighbour telephone authorities, with the notable exception of Germany, to a meeting in Paris in 1923. At this meeting the three suggestions made by Mr Gill were discussed by the French, Belgian, Italian, Swiss, British and Spanish telephone authorities in what was called the *Comité Technique Préliminaire pour la téléphonie à grande distance en Europe*.[6]

The choice of organisation is a rather interesting issue. At the Paris conference all three of Gill's proposals were considered. The first was favoured by the Italians, but rejected as an attempt by the Bell system to take over European long-distance traffic as well as the American. Gill was not only President of the Institution of

---

[5] Gill, F. (1924)

[6] Comité Technique Preliminaire 12/3 1923

42

Electrical Engineers, but also vice-president of International Western Electric, that is, the international branch of the Bell system's manufacturing branch.

The second alternative was also turned down, without much discussion. But clearly the creation of such an international company would mean giving up a bit of national sovereignty over communications policy, and for all the internationalism of the 1920's the participating countries were not prepared to do this.

The third suggestion however, though originally put forward as a plan of study, was adopted. The *Comité Technique Préliminaire* proposed the formation of a permanent advisory committee for international telephony, taking as its task to prepare the organisation of the European international telephony, and to provide for unity of direction in means and ends of international telephony. A part of this would be to centralise technical and statistical information regarding European international telephony. The *Comité* also decided that its proposals should be taken as recommendations of a general character, which the various telephone authorities were advised to follow in their own interest, as well as in the interest of the European network as a whole.[7]

To ensure the continuity of its work the Comité decided to form a permanent subsecretariat which would represent the Comité in between its annual meetings, and prepare the coming conference.[8]

The chosen form was a rather weak organisation, in which participation was voluntary, as was the following of its recommendations. Nevertheless it proved very efficient in opening up the European continent to international telephony, as during the following years most of the other European countries joined.

Having reached decisions on a number of issues regarding technical standards and a plan of international lines found the most urgent to be constructed, the Comité decided to invite other telephone authorities to its next meeting in order to bring all of Europe into the co-operation.[9]

---

[7] Comité Technique Preliminaire 12/3 1923
[8] Ibid.
[9] Ibid.

Table 3.2 Plenary Assemblies of the CCIF

1st	Paris	1924
2nd	Paris	1925
3rd	Paris	1926
4th	Como	1927
5th	Paris	1928
6th	Berlin	1929
7th	Brussels	1930
8th	Paris	1931
9th	Madrid	1932
10th	Budapest	1934
11th	Copenhagen	1936
12th	Cairo	1938
13th	London	1945
14th	Montreux	1946
15th	Paris	1949
16th	Florence	1951
17th	Geneva	1954
18th	Geneva	1956

3.2 Plenary Assemblies of the CCIF

On the basis of the guidelines drawn up at the 1923 meeting in Paris, a new organisation for the co-ordination of international telephony in Europe was created. This gave a new forum for co-operation, and opened a new channel for settling problems of co-ordination. As we shall see, once the channel was established, its capacities gradually expanded, and new areas were brought into the co-operation. The process could be described like if the international co-operation picked up a momentum of its own. In the following section I will describe this increasing momentum by briefly summarising which were the main issues discussed at the Plenary Assemblies up until the Second World War. The purpose of this is to give an idea of the expanding capacities of the CCIF, and show how new fields of international telephone regulation gradually was brought into the co-operation.

3.2.1 Paris 1924

The following year, in April 1924, the six countries forming the Comité Preliminaire were joined by fourteen others[10] at a conference in Paris. At this conference the

---

[10] Namely Denmark, Finland, Yugoslavia, Latvia, Luxemburg, the Netherlands, Norway, Poland, Romania, Sweden, Czechoslovakia, Germany, Hungary and Austria.

44

Comité was given a more solid form; the preliminary organisation that was introduced a year earlier was established permanently, and the new organisation was given the name *Comité Consultatif International des communications téléphoniques à grande distance*, or shorter: CCI.

The conference had been preceded by an intensive work of collecting information from the various telephone authorities about what they felt needed to be co-ordinated and standardised. This resulted in some 50 recommendations on different technical issues, including a plan of the international telephone connections conceived to be the most urgent to construct. Most of the recommendations concerned questions of transmission, quite naturally since the main problem in international telephony at this stage was to build a network with acceptable quality of transmission over long distances.[11]

From the beginning of its existence, the CCI divided the specialised technical work between Committees of Rapporteurs; made up of officials with special competencies in the relevant fields. Originally three such committees were set up, covering the fields of Transmission; Line maintenance and supervision; and Traffic and operations.

3.2.2 Paris 1925

At the second Plenary Assembly in Paris 1925, two related questions of organisation needed to be solved. One concerned the CCI's relations to the League of Nations, and the other the status of the CCI as an international organisation, which up until then had been that of a sort of private, provisional organisation.[12]

According to the peace treaty of Versailles, all new international organisations should automatically be attached to the League of Nations. Germany, which had played a very active role in the CCI but had not yet become a member of the League of Nations, would in this way have been excluded from the Committee's work.[13]

The International Telegraph Union had up to this point showed relatively little interest for the questions of telephony, but as the field grew more important for

---

[11]Heimbürger (1974), pp. 314-315
[12]Chapuis (1976), p. 188
[13]Heimbürger (1974), pp. 315-316

international communications, the Union somehow had to engage itself into these issues. At the same time the CCI had showed considerable success in co-ordinating research and issuing recommendations, which definitely had come to good use in establishing new international circuits.

These two problems were both solved at the 1925 International Telegraph Conference, where it was decided that the CCI would be officially recognised and attached to the International Telegraph Union in Bern. The CCI was still to be independent to some degree though, by being allowed to select its own bureau and decide on its own rules of procedure and methods of work. In this way the CCI was given authorisation as an international body, and was at the same time, through the ITU, affiliated to the League of Nations.[14]

3.2.3 Paris 1926

After having been officially recognised, the CCI at its third Plenary Assembly in Paris 1926 concentrated on establishing its internal organisation and methods of work, as well as its relations to other international bodies. The form of organisation chosen was to remain throughout the existence of the CCIF, and was used as a pattern for the following other consultative committees, i.e. the International Consultative Committees for Telegraph Communications (established in 1925) and for Radio (established in 1927). In this form the CCIF has three main organs: the Plenary Assembly, the Committees of Rapporteurs and the General Secretariat.

The Plenary Assembly is the main deciding body of the CCIF. It is comprised of all member states and its main function is to accept , reject or modify the reports submitted by the Committees of Rapporteurs.[15]

The Committees of Rapporteurs consist of technically skilled officials who in depth study new questions in the area of telephony. Their task is to provide the Plenary Assembly with detailed reports on their findings, and to produce draft recommendations for the Plenary Assembly to decide on. Each Committee is headed by a chairman to direct its work.[16]

---

[14]Chapuis (1976), p. 188
[15] Ibid.
[16] Ibid.

46

The General Secretariat is the administrative body of the organisation. It publishes the reports from the Committees of Rapporteurs and arranges the Plenary Assemblies.[17]

Furthermore the relations of the CCIF to other international organisations were discussed, and technical co-operation established with a wide range of bodies that could possibly concern the field of international telephony. The international bureau of the ITU was invited to attend the Plenary Assemblies and, when considered useful, to participate in the Committees of Rapporteurs.[18]

3.2.4 Como 1927

At the following Plenary Assembly, held in 1927 in Como, a new committee of Rapporteurs was established to deal exclusively with all questions regarding the quality of telephone transmission. As a means of further standardisation in this area, a special laboratory was established, and a reference system which was to function as a standard "metre" of transmission quality, referred to by its French abbreviation SFERT, was installed. This may be seen as one important early step towards widening the CCIF's work to include not only European telephony, as the Plenary Assembly also decided that two identical systems should be installed in New York and Paris. The technical equipment of the laboratories was constructed by the American monopolist on international telephony, AT&T, and donated to the CCIF.[19]

This co-operation between the CCIF and AT&T continued, and at the fifth Plenary Assembly in Paris 1928, two representatives of the latter were present as observers. This resulted in 1930, at the seventh Plenary Assembly in Brussels, in the first extra-European member of the CCIF, when the AT&T joined by officially announcing its membership.[20]

---

[17]Chapuis (1976), p. 188

[18] Ibid.

[19]Ibid., p. 189

[20]Ibid. Worth noting in this context is, however, that the CCIF was concerned with what was known as 'the European system'. This included the telephone system on the European continent and any other who declared that they were part of it. AT&T was part of, and to a large extent constituted, the American system of telephony.

3.2.5 Madrid 1932

The ninth Plenary Assembly of the CCIF took place in Madrid in 1932, and dealt exclusively with questions of operation and tariffs. It was planned to coincide with the International Telegraph and Radiotelegraph Conferences. One of the purposes of the International Telegraph Conference was to revise the section dealing with telephony in the Telegraph Regulations, in order to draw up a set of international telephone regulations, separate from the telegraph regulations. The CCIF had already in 1929 decided to set up a special Committee for revision of the telephone regulations, and the work of this committee was used as a basis for the suggestions made at the Telegraph Conference.

Another issue discussed at the International Telegraph and Radiotelegraph Conferences was the standardisation of the three international consulting committees. The result of this was the bringing of the telegraph and radio committees more in line with the organisation of the CCIF. The three committees were decided to be known as:

- •International Telephone Consultative Committee (CCIF);
- •International Telegraph Consultative Committee (CCIT); and
- •International Radio Consultative Committee (CCIR).[21]

Another change in the CCIF's activities brought about by the International Telegraph Conference was that the Plenary Assemblies should thereafter be held with two year intervals instead of every year. Subsequently the next Plenary Assembly took place in 1934.

3.2.6 1936-39

One important decision of the XIth Plenary Assembly, meeting in Copenhagen in 1936, was to form a "Joint Committee for the General Interconnection Programme in Europe". Already at its first Plenary Assembly in 1924 the CCI had issued a plan of urgent international circuits to be constructed, and this committee's instructions were

---

[21]Chapuis (1976), p. 190

48

to develop this, and to prepare directives for the routing of international telephone calls in Europe.[22]

The last formal Plenary Assembly to be held before the war was in Cairo in 1938. This, like the Madrid meeting, was entirely devoted to questions relating to operation and tariffs, while more technical question were referred to a meeting of the Committees of Rapporteurs, held in Oslo later the same year. By a special procedure the Oslo meeting was empowered to adopt technical recommendations, so that also that meeting had something of the status of a Plenary Assembly. This was the last meeting of this sort to be held before the outbreak of the Second World War.[23]

3.2.7   After the War

During the War the CCIF closed down its business. The hostilities had destroyed much of the European telephone system. During the War transmission techniques had developed rapidly, and the American forces had laid down long distance communication as the front advanced. The CCIF were busy studying and standardising these new technologies into civilian use.[24]

The initiative to revive the ITU after the war was taken by the United States, who invited the other members to a Plenipotentiary Conference in Atlantic City in 1947. That conference showed to be something of a watershed in the history of the CCIF. From having been by and large a European affair, the driving force in the organization now came to be the United States. At the same time the focus of the CCIF shifted to a more international outlook, as the ITU and hence the CCIF became part of the United Nations.

A subtle yet very important change of the International Telecommunications Convention was introduced at the Atlantic City conference. The paragraph dealing with the role of the CCIF was redefined as to say that the terms of reference of the CCIF were "relating to telephony" instead of, as earlier, "relating to *international* telephony". This small change of the wording of the statutes proved to be very

---

[22]Chapuis (1976), p. 192, CCI list of urgent international circuits in Telia's archives Administrativa byrån/Rkonomi- och Kanslibyrån, F IVb: 42:1
[23]Chapuis (1976), pp. 191-192
[24]Chapuis(1976), p. 192

important for the further development of the CCIF. In the following years, and especially after 1956, the CCIF gave more and more attention to questions relating not to international telephony per se, but rather with the operation of national telephone networks.[25] This clearly expanded the capacities of the CCIF to include new areas, and came to some extent to change the agenda of the coming work of the organization.

3.3 Summary

In this chapter we have seen how the European states went about to create an organization to overcome the co-ordination problem of international telephony. The perception of the telephone network as a natural monopoly lead to the opinion that the operation of the European network should be carried out in the form of either a monopoly, or a cartel of national monopolies. In 1923 the cartel arrangement was chosen, and the CCIF set up to provide some unifying force between the national monopolies.

The chosen form of organization was a rather weak one, without any power to enforce its decisions on the member states. Instead the CCIF issued recommendations, hoping that they would be followed in the interest of the national systems as well as that of the whole European network. The initial tasks of the new organization was to deal with the immediate technical problems that had caused its creation. An abundance of recommendations were decided upon regarding transmission, line maintenance, and traffic and operations. The result of these agreements were seen in the mid-1920's as a large number of new lines opened between European countries.

After the first few years of functioning, the CCIF had to turn to problems of its organization and methods of work. It became attached to the International Telecommunications Union, ITU, and through that affiliated to the League of Nations. Its internal procedures of work were fixed in a form where almost all issues were prepared by expert study groups of Rapporteurs, whose reports were then decided on by the Plenary Assembly. A small General Secretariat was established to

---

[25]Chapuis (1976), p. 193

50

keep the work running in between Plenary sessions, and to publish the reports of the Committees of Rapporteurs. After thus consolidating its structure the CCIF gradually expanded its interests to include new fields of international telephony, such as which services should be offered, questions of tariffs, and after the Second World War even questions of the national systems.

The implications of this development for the scope for strategic action on behalf of the member states are mixed. On one hand the lack of power to implement the CCIF's decisions theoretically meant that states who disagreed with any of its decisions could ignore them. On the other, it is true that following the recommendations was in the national interests as well. The expanding capacities of the CCIF made it an ever more important arena for international politics. At the same time the increased number of questions to be dealt with transferred importance from the Plenary Assemblies to the Committees of Rapporteurs, where all the important technical issues were prepared before they came on the Plenary Assembly's agenda.

4. British positions

If we are to study how Britain tried to influence the whole of the European regulatory system with her own preferences, we must begin by establishing which those preferences were. We can be fairly sure that there never existed any pre-defined masterplan of how British representatives in the CCIF would act to further national interests. Nevertheless I will argue in this chapter that by studying the political and institutional setting in which international telephone policy was one part, we can reach conclusions about which interests the British representatives in the CCIF would have wanted to pursue.

As we shall see, the formation of those objectives were a complex process. One that can not be understood properly if we use a state-centred perspective and regard the British 'national interest' as something monolithic. Instead the British objectives grew out of a system of interdependences between a number of different actors at various hierarchic levels; actors who had to co-operate but did not necessarily have the same goals with that co-operation. To explain how this happened, I will start by describing the institutional structure of the British telephone system.

4.1 The structure of the British telephone system

When the telephone was introduced in Britain, very early after its invention, the interest of the Crown in the new means of communication was very limited. The Post Office, having established its rights to monopoly in operating telegraphs through the Telegraph Acts of 1863 and 1868, could arguably be the natural operator of a national telephone system as well. Despite the fact that Victorian Britain was highly pro-market and competition in most fields of the economy, much of the sector of communication was by tradition and long history placed in the hands of the Crown.

The new invention had to show its usefulness as a national system before the Post Office started to get interested, and in that way the market was allowed to carry the initial risks of developing the new systems. That however gave the first telephone companies an advantage in technological know-how, that proved to their advantage when the Post Office took over the regulatory responsibility. This can possibly be also an explanation for the cautiousness, or 'restrictionism' of influential forces within the Post Office and Treasury, after the Post Office had taken over the British telephone system.

It will be argued that due to the hesitant attitude towards the telephone, Britain's telephone network was inferior to that of many of her European counterparts, well into the inter-War era, and that this can explain some of the British strategies in the evolving international co-operation on creating a European telephone network. By studying the institutional structure of the British telecommunications system, we can then see what sort of pressures this structure imposed on the British telephone policy and, ultimately, reach an understanding of what the British 'position' or political goals were in relation to the international telephone service.

4.2   The road to state monopoly 1887-1912

In the period immediately following the introduction of the telephone in Britain, much effort was spent in debating how to deal with the telephone as a new means of communications. The centre of the debate was whether the responsibility for it should be placed in the hands of the state, in the hands of the private entrepreneurs in the market, or something in between the two. After Alexander Graham Bell had invented the first real functional telephone in 1876, he visited Britain in 1877 to try to spread the use of his new invention. To that effect he also approached the Post Office, hoping that they would adopt it and build a network of exchanges. The Engineer-in-Chief, R. S. Culley, advised against such a commitment, on the grounds that the possible future use of such an instrument would be too limited. It has been argued that the view of the Post Office was typical, and that they were, as a rule, negative to any new technology.[1]

Thus disappointed, Bell turned to the private sector instead. This led to the creation of The Telephone Company Ltd in June 1878, as the first telephone operator in Great Britain. This is not to say that the private enterprise solution automatically led to competition in the new industry. First of all the new Telephone Company had the patent rights to the telephone. It could however not control the development of technological substitutes and improvements. One such was the carbon receiver,

---

[1]Pitt, (1980), p. 25

invented by Thomas Alva Edison. This device vastly improved the quality of sound in telephony, and led to a fierce competition between the Telephone Company and the Edison Telephone Company in 1879. The dangers of such competition concerned both parties, and quickly led to an amalgamation of the two into the United Telephone Company. Thus we can find a precedent of the coming structure of the telephone industry in the first year of its existence in Britain.

Gradually the interest of the Post Office for the developing and expanding telephone industry awoke. One of the reasons behind the Post Office's reluctance to involve itself in the telephone business, when prompted by Bell in 1877, was its unhappy experiences from the telegraph earlier in the 19th century.

In the case of the telegraph system, contemporaries began regarding it as a natural monopoly. Given that competition was economically a bad idea for such a system, and that an unregulated private monopoly was politically unacceptable, state ownership started to look like an attractive option in the 1860's. The British system with a number of firms who together held the same market power as a private monopoly, but without the ability to utilise economies of scale and scope, combined the worst of two worlds.

In an effort to address those problems the British telegraph system was integrated into a state monopoly under the Telegraph Acts of 1868 and 1869. The immediate advantages of integrating a number of systems into one single network, and the reduction of tariffs led to dramatic increases in the telegraph traffic after the Post Office's take-over in 1870. The number of telegrams sent in the year ending March 1871 was 9.8 million, compared to the 6.4 million messages sent in 1868.[2]

The economic gains from integration were, however, not as great as was believed. Foreman-Peck and Millward (1994) suggest that the take-over by the Post Office opened the telegraph system to political pressures which both increased costs and decreased the income of the services. The purchase of the private system also proved to be far more expensive than anticipated, with the former private owners selling their stock at inflated prices.[3]

---

[2] Foreman-Peck & Millward (1994), p. 74
[3] Ibid., pp. 74-76, Pitt (1980), p. 26

54

Unwilling to make a similar mistake with the new technology, the Post Office preferred to avoid it in the beginning. However, that very same reason was used as an argument for its later involvement. Having invested so heavily in the telegraph network, the Post Office feared that its revenues from the telegraph would be diminished by a rapid spread of telephony. By buying up private companies and regulating the industry themselves, the Post Office could both protect their investment in the telegraph by restricting the spread of the telephone, and at the same time profit from the new market emerging from it.[4]

The second reason was the perhaps more idealistically expected role of a public service office. In envisaging a national network of telephones, this was seen as requiring a unified system of administration. Considering the perceived "natural monopoly" feature of the telephone business, such a system of administration should be put in the hands of the state. In favour of this argument is the number of technological improvements that had been made to the original telephone. Perhaps most important was the adaptation of the above mentioned carbon receiver. By improving the quality of sound in the transmission of telephone messages, the range of telephone calls could be increased. This was clearly a necessary technological improvement if an inter-urban telephone network was to develop, and, in extension, a national system. It could therefore be argued that only as the telephone could provide a national system of communication it became the business of the Post Office. This was not, however, the grounds on which the involvement of the Post Office was argued.

4.2.1 Legal Post Office monopoly

The method of acquiring such control over the telephone system was to try to include telephony in the sector of communication traditionally being the monopoly of the Crown. The Telegraph Acts of 1863 and 1868 had given such monopoly power over the telegraph system to the Postmaster General. In the Telegraph Act of 1863, the term 'telegraph' is defined as "a wire or wires used for the purpose of telegraphic

---

[4]Pitt, (1980), p. 26

communication...and any apparatus connected therewith..."[5] The Telegraph Act of 1869 then widened the definition of 'telegraph' in these acts to include "...any apparatus for transmitting messages or other communications by means of electric signals...The term 'telegram' shall mean any message or other communication transmitted or intended for transmission by a telegraph."[6] It further stated that "The Postmaster General...shall...have the exclusive privilege of transmitting telegrams within the United Kingdom of Great Britain and Ireland..."[7]

Technically speaking then, it seemed clear that since the telephone was a means for communication through the use of electric signals, it was in the meaning of the Telegraph Acts a telegraph, and should therefore be the monopoly privilege of the Post Office. As we have seen though, the Post Office chose not to exercise that monopoly, but rather left it to the market forces. When they decided to get involved, the claim for this was made on the legal reasons of the Postmaster General's monopoly power. In the case *The Attorney-General v. The Edison Telephone Company of London (Limited)* in 1880 the High Court ruled in favour of the Post Office's complaint.[8] As was later made clear; "For the purpose of the Telegraph Acts a telephone is a telegraph, and a telephonic message is a telegram."[9]

With the legal rights to monopoly thus established, the Post Office did not, however, immediately become the public monopoly operator stipulated by the High Court's interpretation of the Telegraph Acts. The situation was one of conflictual powers. The Postmaster General had the legal right to operate a telephone system within the United Kingdom. The United Telephone Company and its subsidiaries held the patent rights to the Bell and Edison inventions. Pitt (1980) explains the persistence of private telephone operation in Britain by a dualism in the Crown's attitudes towards the telephone system. On the one hand there was an 'expansionist' wing in the Post Office, intending to develop the telephone operations of the Department into a coherent state-owned monopoly operator. On the other hand were

---

[5]Telegraph Act 1863, An Act to regulate the Exercise of Powers under Special Acts for the Construction and Maintenance of Telegraphs. (26 & 27 Vict. c. 112), s. 3
[6]Telegraph Act 1869, An Act to alter and amend "The Telegraph Act 1868." (32 &33 Vict. c. 73), s.3
[7]Ibid. s.4
[8]Queen's Bench, C. P., and Ex. Divisions. Dec. 20 1880.

the 'restrictionist' forces, most importantly within Treasury, whose view was that the role of the Post Office's telephone installations mainly was to act as a counter-balance to the power of the private companies. These companies would then operate under licenses from the Post Office.[10]

The first negotiations of such a licence began in 1881. The result was a licence for metropolitan operation of the United Company, granting the Post Office 10% of the gross revenue from business carried out within an area in central London. Another requirement was that the United Company would provide the Post Office with telephones in order for it to establish exchanges of its own in a number of other towns in the United Kingdom. The Postmaster General did, however, further announce that he was willing to grant further licences to the United Company or other companies to establish exchanges in other parts of the country, given that both the Post Office and the public could be given adequate protection.

This settlement did not satisfy any of the partners. The requirement of the United Company to hand over telephones to the Department effectively was an infringement on its patent rights. On the other hand it was argued by the 'expansionist' forces within the Post Office that the assigning of state monopoly powers to private concerns was both unlawful and wasteful of public resources. The competitive power of the Post Office remained weak. Treasury did not provide the funding or power to establish a competitive state-owned system, and still in 1892 the Post Office operated a mere 35 exchanges with a total number of subscribers at 4691.[11]

The favour of Government was with the 'restrictionists', and in 1884 the licensing policy was extended, so that all licensees could operate freely within the country, paying a 10% royalty to the Post Office. This meant that the earlier concerns of protecting the public and Post Office interests by restricting where private companies could operate were abandoned. The effect of this was not, however, increased competition. Rather the opposite. The three major private telephone

---

[9]Halsbury's Laws of England, Telegraphs and Telephones, part V, para. 38, p. 30,

[10] Another example is provided by the Danish telephone system. There the public monopoly rights to telephone operation were established in 1896. Just like in Britain however, the expenses to nationalise the existing private networks were judged prohibitive, and the private companies continued to operate under licenses from the Ministry of Public Works. Israelsen (1992), pp. 245-259

[11]Robertson (1947), cited in Pitt (1980), p. 29

operators that were active towards the end of the 1880's amalgamated in 1889 into the National Telephone Company (NTC). Thus, in the absence of a national, state owned monopoly operator envisaged in the High Court ruling of 1880, what had evolved was not a system of a competitive market, but a private monopoly.

The explanation that immediately springs to mind for this is of course the inherent tendencies towards natural monopoly in the telephone business. First technology had made it possible to make telephone calls over long distances. By the extension of the rights of licensees to conduct business wherever they liked, the regulatory barriers to operating inter-urban networks were removed. This provided the opportunity to benefit from the positive network externalities involved in that kind of network systems. Since the benefit from being part of a network increases with the number of additional members, this clearly gives large operators an advantage over small, and hence the inherent natural monopoly tendencies.[12]

There were however additional forces, inclining the major private operators towards amalgamation. The master telephone patents were about to expire in 1891 and '92, and amalgamation then seemed a rational strategy for pre-empting the expected competition from new operators. Secondly the licence agreement of 1884 had given the Post Office the right to purchase the assets of the licensed companies at certain specified dates. It was reasoned that amalgamation would strengthen the bargaining positions of the licensees, in the event of such a purchase, in order to maximise the value of their stock.[13]

As predicted by the NTC, the expiry of the master patents did indeed lead to a number of new competitors, mainly on the local level. These often took the form of co-operatives or mutual benefit companies, providing low-cost telephone services to their shareholders. One force restricting the NTC's ability to compete with these new entrants was the persisting threat of being bought up by the Post Office at the option purchase dates. That possibility acted as a disincentive to invest. The NTC seems to

---

[12]At some point, of course, these positive externalities start decreasing again. To the average telephone subscriber in the UK today, the marginal additional value of a new telephone extension in, say, rural China is limited.

[13]Pitt (1980), pp. 30 f.

have had a problem of raising capital, which in turn led to the company retaining outdated technology.

As predicted by economic theory however, the new entrants in the telephone market could not compete successfully with the NTC in the long run. The logic of the situation made it cheaper for one big company to supply a service than it was for two smaller ones to do the same. In 1892 this led the main competitor of the NTC, the New Telephone Company, to suggest a merger with the NTC.

4.2.2 Trunk line operation

That merger provided something of a turning point in the Crown's relations towards operating the national telephone network. In the agreement permitting the merger between the two in the Telegraph Act of 1892, the NTC agreed to sell all its existing trunk lines and to refrain from constructing any new trunk lines in the future. It also meant that the Post Office came to have the full responsibility for the international traffic, still in its very first, experimental stages.

At the time this was perhaps not as dramatic as it sounds. The operation of trunk lines was the least profitable sector of telephone operation, and the NTC had previously complained that they had to bind their capital into these investments while their competitors could concentrate wholly on the more lucrative local exchange market. From the point of view of the Post Office the take-over of the trunk system was a means of assuring that the decreased revenue from the state-owned telegraph system would be compensated for by the telephone incomes. The Postmaster General expressed the state-financial importance of the trunk system: "The safeguard of the taxpayer will be our possession of the trunk wires..."[14] The take-over in reality also gave the Post Office a new position as a substantial operator, and this gave the department a new role in the telephone business.

This new position was built upon as the next option purchase date drew nearer in 1897. The public and press opinion, earlier rather hostile towards any attempts from the Post Office to restrict the private operators' freedom to provide telephone services, had now turned on the monopoly power of the NTC instead. Likewise the

---

[14]Meyer (1907), p.46

mood of the Government was now more inclined to accept the long held view of the expansionists in the department. In 1905 a select committee was appointed to study the possibilities of a Post Office take-over of the NTC. The committee recommended that the state should purchase the assets of the NTC on the 31 December 1911. After a long process of valuation and arbitration this was done, and the British telephone system became the state owned monopoly one could have expected it to be from its beginnings.

As we have seen in this part, the British telephone network first evolved in private hands. This can not primarily be explained by a desire on behalf of the state to keep out of operating the communications industries, but rather an unwillingness to make a risky investment in an industry that yet had to prove its importance. That underlying explanation is useful to keep in mind when we proceed to investigate further the forces that lay behind the British positions in the international telephone work.

When the decision was made to take over the national telephone network, again, this was not due to some optimistic view of the telephone as the means of communication for a new era. Rather the take-over was caused by fears that the telephone would cause losses to the Post Office's lucrative telegraph monopoly. Again, the hesitant attitude towards the new technology, which so clearly marked the difference between the Post Office officials and the 'telephone men' of the private operators is worth noting, as is the Post Office's view of the telephone system as primarily a means of making revenue or at least cutting losses on their telegraph investment.

4.3 Forces shaping the British position, 1912-1939

Having given a brief account for how the British national telephone network was built up and eventually taken over by the Post Office and turned into a state monopoly, we will now turn our attention to how this affected the British positions in the international co-operation. First of all it must be noted that the international co-operation within Europe was based largely on state-owned monopoly operators. In a way it could be argued that the existence of a national monopoly in operation was a necessary precondition for the existence of a single 'British position'. In the

60

hypothetical case of the British telephone system being run by a number of competing companies, it can be assumed that these would not have had identical goals, but advocated their own interests individually, rather than some common British interest.

When, in the early 1920's, the CCIF was formed, this was done as we have seen, through the contacts between the various telegraph and telephone state Administrations, and as the organisation grew later in the century to include new members who did not have a state monopoly, this posed new problems as to who should be the 'legitimate representative' of those member states. Precisely that problem had occurred at the formation of the International Telegraph Union in 1865, when Britain was not invited to join the Union until the British telegraph system had been monopolised by the Post Office.[15]

However, as the British telephone system came under the control of the Post Office, it becomes meaningful to try to identify a single British position, or a set of preferences, which they tried to put forward in the international telephone work. In this section we shall therefore study a number of factors in the national telephone system that had some influence over the British preferences internationally.

4.3.1 Technological backwardness

When the agreement was reached for the Post Office to buy up the existing plant of the NTC, the price was to be fixed by arbitration. This further provided a disincentive for the NTC to invest in the period between the reaching of the agreement in 1905, and the actual take-over in 1912, since the outcome of the arbitration was uncertain, and the company ran the risk of not recovering their costs. The result of this was not only to make the company more vulnerable to competition from new entrants with no sunk costs in existing technology, but also arguably led to a degree of backwardness of the whole British telephone system and hindered the more general spread of telephony. After the Post Office eventually took over the whole system it had greatly

---

[15]Grimm (1972), p. 15; As a matter of interest, the same problems reappeared in the 1990's as many of the European telephone markets started to become liberalised and opened to competition. Once more much concern was made over the problem of who should be the 'legitimate representative' in international telecommunications work when the once monopolist state Administrations suddenly

deteriorated, and the capacity of the British telephone system was quite simply unable to meet the demand.[16]

The spread of the telephone to large groups of people was very limited, and the number of telephones per 1000 inhabitants was still, in 1929, approximately half of that of Sweden and one fourth of that of the United States.[17] One historian even goes so far as to discuss the failure of the telephone to make an impression on everyday life in the early years of British telephony.[18]

Table 4.1 Telephones per 100 inhabitants, selected countries

	1913	1932
United States	9.1	14.3
Canada	5.6	12.09
New Zealand	4.0	10.1
Denmark	4.2	9.8
Sweden	3.9	9.3
Switzerland	2.3	8.5
Australia	2.6	7.4
Norway	3.1	7.0
Great Britain	1.6	4.63
Germany	1.9	4.57
France	0.7	3.0

Data from Foreman-Peck (1992), p. 173

The spread of the telephone into everyday life of the British did not automatically speed up after the insecurities of the future ownership of the telephone system had disappeared (see table 4.1). If the NTC had been reluctant in investing in new technology before because of fears that their stock would be taken over by the state, the state as owner were just as reluctant because of fears that it would not be profitable in the short run to undertake such investments. This especially held true for the expansion of the telephone network into rural areas. In the years immediately

---

became one among several competitors on the national markets. See for instance SOU 1992:70, chapter 7.4, pp. 155-164

[16]Sir Evelyn Murray, K. C. B., Secretary to the Post Office, before the Select Committee on the Telephone Service, 6 April, 1921. Report of the Select Committee on the Telephone Service, 1921-2; *Parliamentary Papers 1921* vii, (191), 431, pp. 443-4

[17]Sellars (1933)

[18]Perry (1977)

following the Post Office's take-over of the NTC network there was a constant, but very small, yearly increase in the number of stations. The demand for telephony was bigger than the supply, creating substantial waiting times for telephones, and an increasing lack of popularity for the Post Office. This time the possibility of trying to blame the private operators was not available either.

During the years of the First World War not much criticism of the public telephone network was heard, but following the peace the problems of not meeting the demand started again. This time in addition to the cautiousness of Government to invest in something as risky as a telephone network, there was also the substantial strain on state finances supplied by the 1920's depression. The Treasury under the Conservative government had no great belief in the idea of investing in telephone lines to boost domestic demand. Overall, the belief in public works was very limited, and in the cases where such were undertaken it was in rather more 'safe' areas such as road building.

The rising dissatisfaction with the state of the nation's telephone system led to the setting up of a select committee on the telephone service in 1920. The main findings of the committee was that the Post Office suffered from an inadequate bureaucratic structure. In order to successfully run such a 'commercial' feature of government business such as actually operating a telephone network, the management of the department would have to rely on specialist and technical staff rather than civil servant bureaucrats.[19] The examples of the United States and continental Europe were cited, where senior administrators had to work their way up through the organisation in order to learn their work. This was contrasted against the British Post Office where "...gentlemen who, having passed an examination, are put straight away in the central organisation with no opportunity of learning the work except by correspondence."[20] However, this view was hardly uncontroversial with the Treasury, and the debate on the future of the Post Office continued throughout the 1920's and well into the 1930's.

---

[19]Report from the Select Committee on the Telephone Service, 1922 (197) vi
[20]Lt.-Col. Walter Alfred John O'Meara, C.M.G., representing the London Telephone Advisory Committee, before the Select Committee on the Telephone Service, 11 July, 1921; *Parliamentary Papers 1921* vii (191) 431, § 6506

In summary then, the British national telephone system was at the time of the Post Office's take-over in poor shape. Furthermore the state as the new owner of the system was not willing or able to invest heavily in the network in order to increase its standard. The relatively low standard of the British national system of course had some influence in determining the priorities of the Post Office regarding the international network, as we shall see. Also the fact that the spread of the telephone to the British public was comparatively low, meant that investment in an improved international service perhaps was not a top priority in the national communications policy.

4.3.2 Pricing policy

The deficiencies of the British telephone system were also related to the pricing policy. In the United States, services were technically superior and the telephone lines were less busy. Often no waiting time at all was necessary for placing a telephone call. The rates were also substantially higher. This was a principal difference between the American and the European systems, which all seem to have been more or less following a low-price, lower quality policy. The American long-distance rates were considerably higher in the United States than in Europe, both in absolute terms and as a proportion of the local rates. The reason for this was due to the fact that the European trunk systems were dimensioned for low-traffic periods, whereas AT&T's long distance system held excess capacity for large parts of the day. In accordance, long waiting times to put a call through, which was a typical phenomenon of all European long-distance telephony, were almost unheard of in America.[21]

With large sums taken out of the state budget that had to be invested in the setting up of a state-owned telephone system, one of the political pressures arising was that the long-distance service should be cheap. The trunk system, as operated by the Post Office, had by the time of the inquiries of the Select Committee on the Telephone Service run an annual deficit. With those low rates and the subsequently

---

[21]Holcombe (1911), pp. 413-417

relatively low level of investment in trunk lines, those lines were inevitably congested.[22]

The 1922 Select Committee on the Telephone Service mentions the accounting methods of the Post Office as one factor contributing to the underinvestment in the trunk line system. Whereas a private company would keep a capital account, from which capital investments were charged, all the surplus profits from the Post Office went to the Treasury; and all losses were also met by the Treasury. That, concluded the Committee, created a danger "...that such expenditure will be regarded as imposing a burden on the taxpayer, rather than as a stimulus to trade, and a remunerative investment for the telephone department."[23]

One feature of the early days of telephony that persisted for a long time in most of the European systems was the flat rate charge. In most cases, the early telephone systems were private lines, or at least consisted of a small number of subscribers. So was also the amount of traffic that passed over the system. In those cases the cost of the service was chiefly made up of the cost of the capital investment, depreciation of plant, and maintenance costs. It then makes sense that these early systems charged their users a flat rate, irrespective of how much they used them.

As the systems grew bigger and both the number of subscribers and the number of calls increased, the composition of costs changed. In addition to the earlier costs of the telephone service the cost of operation of the system became a much larger share of total costs. The operation of telephone exchanges meant that the costs of operating the system increased with the amount of traffic conducted in it.

This new feature of the telephone industry led to much debate in the early years of the 20th century. Large telephone users naturally saw that it was in their interest to keep the flat rate, and in many cases the European administrations kept that pricing policy long after it had been abolished in the more advanced market of the United States. This was, by propagators for the private telephone industry, taken as evidence of the deficiencies of state-run telephone services. The state-owned monopolies were not run according to business principles, but were open to political pressures which

---

[22]Webb (1911), pp. 46-47
[23]Report from the Select Committee on the Telephone Service, 1922 (197) vi, § 23

hindered the development of telephony. The British Chamber of Commerce was one of the most persistent opponents to any introduction of a measured system of charging.[24]

The effect of maintaining the flat rate was of course that large users were subsidised by smaller ones, and that many would-be small users refrained from becoming subscribers due to the high charge. This meant that the number of subscribers remained relatively low. By the principle of positive network externalities mentioned earlier, that also meant that the value of the telephone service to the already existing users was kept at an unnecessarily low level.

In Britain the message rate, where each subscriber was charged according to the number of calls he made, was introduced on April 1, 1921. At the same time, the charges for trunk line calls were changed. Since the rules for calculating the charges were completely different from the previous system, straight comparisons of the price level before and after the change are difficult to make. The Post Office however estimated the charges for trunk calls to have increased by approximately 50%.

In summary then, the effect of the Post Office's pricing policy on the telephone development in Britain was firstly, through the flat rate, that large users were subsidised by smaller ones. At the same time the low level of the rates and the accounting practices of the Post Office meant that little new investment went into the trunk line system. The result was a trunk service of rather poor quality where long waiting times were the normal situation. One way of dealing with that problem was to raise prices and thus reducing traffic, as was done in 1921. The immediate result of that change in the charging system however, was to increase the price dramatically for a not greatly changed service. The British trunk-line system was still, by international comparisons, relatively old-fashioned.

4.3.3 Insufficient trunk-line capacity

The other factor that kept the number of subscribers down was the insufficient trunk line capacity. Once more compared to the United States, the quality of long-distance telephone service was substantially lower in Europe. In 1910, the waiting time for

---

[24]Baldwin (1925), ch. XXII; Report from the Select Committee on the Telephone Service, *Parliamentary Papers 1922* vi (197), § 41

long-distance calls in the American Bell system was described as normally being "a few minutes...and it is rare for a call between any two of the large cities to be delayed over ten minutes."[25] In Europe at the same time, there would at almost any time of the day be a long line of people waiting for a call to be put through on the long-distance lines, and some calls would have to wait for several hours. This clearly reduced the value of the service, and therefore also the spread of using the telephone.

The strongest opponents to the low-price low-quality policy of the European state telephone administrations were the technical men of the telephone industry. Frank Gill, Engineer-in-Chief of the National Telephone Company and President of the Institution of Electrical Engineers expressed it rather plainly when speaking of the creation of reliable international telephone services in Europe: "...(T)he price is a secondary matter, the really important things are accuracy, speed and quality."[26]

This, then, seems to express one of the primary differences between the AT&T-operated American system, and the state run European telephone networks. The 'expansionism' of the forces within the British telephone administration who wanted a rapid development and improvement of the national long distance service, also implied heavy investment in the system.

The strategy of high quality and high rates did apparently work well in the United States, where AT&T operated their long distance monopoly. In Europe however, there was less support for that type of policy. In part that could be explained by the fact that the operating of the telephone systems was in the hands of the states. By standard public choice assumptions such an arrangement would open the system to political pressures, and that, in turn, would mean that the system was run sub-optimally. The extension of that argument would be that cheap rates dictated by the political unwillingness to use tax payer's money to build up an exclusive and expensive system. That argument would however have less validity if it could be

---

[25]Webb (1911), p. 47

[26]Gill, F. Foreword to Baldwin (1925), p. vi; Also the 1922 Select Committee on the Telephone Service concluded that the British long distance service was of inferior quality to that of the United States, and that the demand for trunk service above all was created by the provision of a reliable service. See Report from the Select Committee on the Telephone Service 1922 (197), § 32. At the same time, however, the report argued for lower rates in order to spread the use of telephones to a wider range of the public. See paras. 21, 22, 28, 48.

shown that the telephone system could be operated with a profit; in which case the public choice logic implies that the political pressures and the business logic would converge. That convergence could of course be disturbed though, for instance by such factors as the accounting principles of the Post Offices, which failed to show investments in the trunk-line system as assets.

Another explanation that focuses on the supply side is that there were vast economical differences between the wealthy United States and the war-torn European economies. The capital needed for a heavy investment program in the European telephone systems either was not available, or would have crowded out other investments, which were judged more pressing.

Also the demand sides of the economies were of course greatly different. Firstly, the overall national wealth of the American economy was higher than that of her European counterparts. Europe was of course damaged by the War. Aldcroft (1993) mentions as a crude figure that the War caused an eight year set-back to the growth of production in Europe.[27] At the same time other countries could benefit from the increased demand for goods and Europe's inability to supply those markets. In the inter-War years Europe never managed to regain her former economic power status. One example of that can be given by the allocation of world trade before and after the First World War.

Table 4.2 Percentage of world trade for certain regions

	1913	1920
The Americas	22.4	32.1
Europe + USSR	58.4	49.2

Data from Aldcroft (1993), p. 14

Still, it must be remembered that Europe's decline was above all relative, not absolute. Despite the mass unemployment, instability and decline in international trade, real incomes rose quite considerably in most countries during the inter-War years. Foreman-Peck (1995) shows that the annual average growth rate of real gross

---

[27]Aldcroft (1993), p. 12

national product per head between 1913 and 1950 was 0.7% in Germany, 0.8% in the United Kingdom, and 1.5% in the United States.[28]

What is important in the argument here however, is the different levels of wealth in Europe and the United States respectively. The GDP-level per capita was substantially higher in the United States than in Europe. Furthermore the real wages differed greatly between Europe and America.

Table 4.3 Index numbers of the relative levels of real wages in the large towns of different countries in July 1930

(Base: Great Britain=100)

Country	Real Wages
United States	190
Canada	155
Denmark	113
Sweden	109
Great Britain	100
Irish Free State	93
Netherlands	82
Germany	73
Poland	61
Austria	48
Yugoslavia	45
Spain	40
Italy	39

Data from Hansen (1932), p. 255

That difference would also mean a greater demand in the United States for an exclusive and expensive item as long-distance telephony. Furthermore the size of the American market would mean not only a greater market over which to spread investment costs, but also a greater number of users demanding a high quality service, where, in the already quoted words, the price was a secondary matter.

4.3.4 Political distrust

When the Post Office took over the assets of the National Telephone Company, on the 31 December 1911, they had neither staff nor experience to run a national telephone system. This meant that much of the NTC staff so to speak came into the

---

[28]Foreman-Peck (1995), p. 184

bargain as well, by simply changing employer. The relations between the Post Office and the old NTC-staff were in many ways under certain amounts of stress. Frank Gill, Chief Engineer of the NTC and notable figure in international telephony issues, did not join the Post Office, and many who did were dissatisfied with the conditions of service. According to one official, even in 1923, 11 years after the take-over, "...the ex-N.T.C. men were still distinguishable, and that not only by the possession of the N.T.C. pencil, a prised souvenir of their former allegiance."[29]

One of the grounds for the Treasury's suspicion of the more technically skilled staff of the telephone branch of the Post Office, might have been that many of them were actually ex-NTC staff, who had continued doing rather the same work as before, though under a new employer.

A brief characterisation of the Post Office during the first decades of the twentieth century, then, suggests that the organisation was highly centralised and top heavy. The proportion of engineers and technically skilled staff was relatively low compared to other European Telephone Administrations. There was also a tradition of distrust between the Treasury and the National Telephone Company, which carried on after the taking over of all the NTC's plant by the Post Office. The lack of decentralisation meant that staff levels were high compared to other State Telephone operators, even if that is difficult to quantify.[30]

The telephone network was, at the time of the take over, in a rather poor condition. Due to the earlier disincentives to invest in new plant for the NTC and the 'restrictionist' policies both within the Post Office and Treasury the network was both technically obsolete and underdimensioned. The shortages of materials, capital and staff because of the First World War meant that these deficiencies could not have been met until the 1920's, but even then the process was slow.

4.3.5 Conflicting goals

One factor that must be considered whenever discussing the British position in international telephone bodies was the fact that even if the British representatives in these fora, i.e. the Post Office, would have been keen to develop the telephone

---

[29]Ray (1954)

network as quickly and thoroughly as possible, the national position ultimately taken would have to be co-ordinated with other political goals. Naturally the Post Office had an important influence in deciding on British positions in the actual dealing with the other Administrations. Those positions however had to be accommodated to an overall communications policy, and that communications policy in turn had to be accommodated to other political goals.

It is therefore not possible to regard the telephone policy in isolation from other fields. Instead we must try to find areas where there were goal conflicts between the expansion of the European international telephone service and other political aims, and discuss how these conflicts did affect the chosen British positions or objectives.

The perhaps most striking feature of the British inter-War economy is the poor shape it was in. In a tightly squeezed economy, the opportunity for heavy investment in the telephone system was very limited. In that situation the Treasury did not share the view of the most enthusiastic telephone propagators of a well-developed telephone network as a critical factor in economic recovery and growth.

The case of conflicting goals between the Post Office and Government can also be found in other fields. The technological limitations of long-distance telephony meant that cable telephony from Britain was limited to the European continent. For a number of reasons however, there was a strong concern on behalf of Britain to keep closely in contact with other parts of the world. Economically, politically and strategically, Britain's position as a major World Power was dependent on her close links with the Empire; and also on her relationship with the United States.

An overall communications policy would therefore have to ensure good communications with those important areas of the world. In that light it is also understandable that the enthusiasm for a continental telephone network was, if not lukewarm then at least less marked, on behalf of the Government.

One of the results of the reorganisation of the Post Office in the 1920's, was to separate it further away from the political influence of the Treasury. This was one of the intentions of the reform, but the consequences of such a separation were probably not anticipated. By letting the 'technical men' of telephony have a greater influence

[30]Ibid.

over the operation of the telephone system, the whole of the Telephone department became more 'expansionist'. Many of the views earlier expressed by the most enthusiastic of proponents for the spread of the telephone now became those of the Post Office. In this way the Post Office also sought new methods of achieving the goal of an expanded and improved telephone system.

The setting up of a co-operation with other national telephone Administrations with the expressed goal of improving and expanding the international service in Europe was one step towards that. In this way the Post Office formed a coalition with the foreign Administrations, as well as with the telephone industry, to rally some political pressure to ascertain the desirability of a well-functioning international service in Europe. This is further illustrated by the fact that representatives from the telephone manufacturing industry were invited to participate in the meetings of the CCIF and its permanent commission. The social network between delegates of the various European Administrations that began to take form after 1923, also included representatives from the industry.

The effect of this change on behalf of the British Post Office, was that it made more clear the division between the 'expansionist' telephone officials and the 'restrictionist' Treasury. The restrictive influences were still very apparent through the political procedures that shaped the British position in the international agreements. The Post Office in itself though, became a more coherent body in its attitudes towards the expansion of telephone services.

Thus we can see the development of what Marin (1990) calls 'antagonistic co-operation'. The British position was reached through the co-operation between the telephone specialists and the politicians. These had non-converging, or even conflicting, interests; but still they had to reach a position. That process can not be simply described as an exchange situation between the two, where they entered into some symmetrical bargaining of exchanging preferences or political assets of some sort. Instead a whole network of actors were interlinked, where the exchange processes could be asymmetrical and complex. Marin calls this sort of exchange process a Generalised Political Exchange.

72

An example of this is how the Treasury and Cabinet, who had the last word in any plans that included some form of public expenditure, left the negotiations with the foreign Administrations to the Post Office, simply because the Post Office had the technical expertise needed. The Post Office in its turn could not make direct demands to the Cabinet. Instead they formed a coalition with the Foreign Administrations and the manufacturing industry, in order to give weight to the claims of the importance of a well-functioning international service in Europe. The immediate effect of the work of the CCIF was to open telephone lines to other countries in Europe. This in turn, it was argued[31], would create a demand for the international services. Such a demand would then work as a political pressure for the expansion and improvement of the telephone system.

4.4 The British international service after 1922

In 1922 the first Anglo-Dutch service was opened. It immediately proved insufficient for meeting the demand for it, and a second submarine cable between Great Britain and the Netherlands was laid out in 1924. Through the lines to the Netherlands telephone contacts with Germany opened in 1926. At first only with service to Berlin, Frankfurt, Hamburg and Cologne, but in 1927 extended to the whole of Germany. In 1927 this was followed by further extensions to the Scandinavian countries; Sweden, Denmark and Norway, and to Vienna, with Budapest and Prague following suit early in 1928. That year the rates for calls to the newly opened lines to Germany and beyond were reduced by some 20-25%.[32]

The expansion of the British international telephone service continued in 1928 with the opening of the Anglo-Spanish and Anglo-Italian services, in 1929 contact was established with Finland and Poland and in 1930 with the Baltic states Estonia, Latvia and Lithuania. This remarkable expansion of the international service meant that in less than ten years Britain had established telephone links to most of Europe, and the direct, or 'through', lines to the continent had risen to 107 in May 1931.[33]

---

[31]for instance by Gill (1925) and Holcombe (1911)
[32]Post Office (1931)
[33]Ibid.

The distribution of traffic in Europe was heavily biased towards the larger neighbouring countries. In the year ended 31 March 1933, the distribution of incoming and outgoing international calls was the following:

Figure 4.1 Proportionate distribution of Anglo-Continental traffic 1932-33

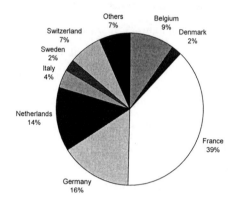

Data from Post Office (1934b)

Another important feature in the development of international telephony was the beginning of wireless telephony. This of course opened up possibilities for telephone contact between geographically very distant areas. Thus, in 1927, the first radiotelephone service between London and New York was opened, and was gradually extended to all of Great Britain and the United States, followed by Cuba and Canada. The year after, the transatlantic service was extended to Mexico, followed by Australia in 1930. This new feature of international telephony then, put Britain in telephone contact with the largest part of the Empire, and in practice provided a link independent of the European cable network. The European system was however being connected to the wireless telephone system, so that subscribers all over Europe could effectively be put in contact with distant parts of the world.[34]

In terms of number of calls however, the radiotelephone service remained a fairly small proportion of the total number of international calls. Nevertheless they were important in providing the opportunity for intercontinental communication (see table 4.4).

---

[34] Post Office (1931)

74

Table 4.4 Number of international calls originating in Britain, 1927-1939

Year	Continental telephone calls	Radiotelephone calls
1927	283121	250
1928	338377	1243
1929	451099	3205
1930	532026	3951
1931	540576	4461
1932	586970	5138
1933	588585	5737
1934	664878	8551
1935	692677	9678
1936	784379	11875
1937	894723	17861
1938	942993	18873
1939	984043	17658

Data from Post Office Telecommunications Statistics, 1949, p. 21

Another medium for intercontinental communications was of course the telegraph. Taking 1939 as a year for comparison, the number of telegrams handled between Britain and the rest of the world was over 7.8 million, and of these approximately 3.1 million were outside Europe.[35] Thus, in terms of messages exchanged, the telegraph was still a more important means of communication internationally up until the Second World War.

Such comparisons must however be made extremely cautiously, if at all, since the telephone and the telegraph provide two qualitatively different services. Given the indirect form of communication provided by exchanging telegrams, that presumably will also affect the content of the messages exchanged.

4.5 British objectives

From the background set out in this chapter, I conclude that as far as we can talk of a British position, or persistent features in British telephone policy towards her European counterparts, it is primarily marked by three such features. One regards the economics of telephony and the insistence that the telephone investments should bear their own costs.

---

[35] Post Office Telecommunications Statistics, 1949, p. 30

The second has to do with the ideal of a free market for telephony, something which involved rational profit maximising in the neo-classical sense as well as more ideological explanations; the notion in British administration that independence for producers to choose freely how they wanted to operate was something good per se. This is a strong political, or even cultural, consistency in British politics according to Dobbin (1994), who sees it as one important factor in making British railway politics different from those of France and the United States.

The third regards Britain's real or perceived place in the world, which resulted in a more international, or less Eurocentric outlook on communications than the other European players in the CCIF.

4.5.1 Business operation of the Post Office

Through the practice of having all agreements with foreign Administrations being cleared with the Foreign Office and Treasury, a certain amount of political influence was always present in the making of those agreements. Given the 'restrictionist' tendencies of, above all, the Treasury, this influenced the British position in the period up to at least the 1930's, making it rather cautious regarding expenditure on new circuits. The comparatively poor state of the domestic telephone network also meant that the Post Office were unwilling to support any international agreements that included heavy investment in the national networks as well.

During the 1920's the telephone operation branch of the Post Office was gradually reorganised, partly as a result of the 1921/22 Select Committee on the Telephone Service. The object of this change was to make the operation more 'commercial' in style. In the process the attitudes towards expanding the international service also gradually changed. The 'technical men' of the telephone service started having a greater influence over the operation of the telephone system, and the Post Office in itself became more of an 'expansionist' body, even if the influence of the Treasury was still great.

The outcome of the commercialisation of the Post Office was to further emphasise the necessity for the telephone branch of the Post Office to be run on business-like principles. Given the reluctance of Treasury to spend taxpayer's money on telephone investment, the telephone operation had to be shown to be good business in itself. The re-organisation of the Post Office and the change of its accounting methods were done to ensure this.

As a consequence, the British views did not always converge with those of the continental Administrations which rather regarded their telephone systems as national infrastructural investments, and saw it as something inherently good to make available cheap telephone communications. Especially regarding international rates, the British were seen as obstructing the beneficial spread of the telephone by insisting on high rates.

4.5.2 Free market policy

As a consequence of the business-like running of the Post Office's telephone branch, the British telephone policy regarding the international network differed from that of many of her continental counterparts. For want of a better expression the British stance could be described as a 'free market policy', which in practice implied favouring a flexible system of prices or rates and as far as possible competition within the international European system. This contrasted sharply against the ideals expressed by many of the other European Administrations, most markedly the German, who as far as possible were aiming at establishing a unified, standardised European network.

This policy at a number of times led to conflict between the Post Office and the continental Administrations, especially in establishing rates for international lines. The continental idea of having a set, uniform rate, calculated from the distance of the communications clashed with the British desire to price each connection individually according to the demand for it.

Furthermore the British Administration envisaged a development of a European market where the various national operators effectively competed with each other for traffic. This in practice was something quite foreign to the European system and to the CCIF, which in effect worked as a cartel in dividing the European market into a number of small, local monopolies.

The motivations behind this free-market policy of the Post Office can in part be found in Britain's geographical location on the fringe of Europe. That naturally meant that, with the exception of international traffic to and from Ireland, there was no transit traffic through Britain. Since competition between the national Administrations, which in fact were local monopolies, only could take the form of alternative routing of calls over different countries, Britain had nothing to lose from a competitive system, but lots of things to win on lower rates on behalf of the other European countries.

In addition to that strictly maximising, neo-classically rational explanation, there is also something to be said for more sociological or cognitive explanations. As Frank Dobbin (1994) has pointed out, the national culture in Britain was one of endorsing private enterprise. This is something deeper than a calculus of benefit, but actually has to do with cognitive structures or, in slightly different terms, national culture.

According to Dobbin, national traditions make an impact on the cognitive systems of people, the way they think. They create different ideas about which way is the best to solve certain problems. History has produced distinct ideas about order and rationality in different nations, and these ideas form the base on which industrial policies are built. His argument is based on the observation that societies tend to approach new problems with tried strategies. A history of applying a certain set of solutions tends to direct people's thoughts of what actually constitutes a 'good solution'. In Britain the official attitude towards what constituted a good answer to economic problems was to give entrepreneurs the greatest independence from interference of both politics and markets.[36]

4.5.3 Extra-European outlook

Following the setting up of the CCIF, the British administration started to co-operate more closely with its foreign counterparts than before. During the 1920's the international service from Britain was extended to a number of new countries. One of the primary differences between the British Administration and the others though, was the marked difference in 'Eurocentricity'.

To the British Administration the developments of wireless telephony was highly important, since it meant that contact could be established with the dominions

---

[36] Dobbin (1994)

and the United States. The British delegations also displayed a greater interest in creating world standards, rather than European. In many cases the position of the British Administration was that of harmonising the European system to the American Bell System.

The notion of an 'extra-European' outlook of the British Administration in terms of its positions visavi the CCIF might sound preposterously Eurocentric. Arguably the British outlook on the world had been decidedly extra-European for a very long time, and thus this was only natural. Seen from the perspective of the developing European telephone network however, it is an interesting fact that the British position in this respect stood out from that of her partners in the European co-operation.

Earlier studies of the international telecommunications organisations have often focused on the specialist, technical, non-political function of these bodies. In relating to the work of, for instance Codding (1952), the British preferences and strategies become incomprehensible unless one tries to relate the telecommunications policy to Britain's other international relations. This calls for a study of British foreign and security policy between the wars.

4.6   British defence policy between the two World Wars

By the end of the First World War, the British Empire can be said to have stood at its height. Britain and her Allies had won the War. The Dominions had taken a most active part and paid a high price, and their presence at the peace negotiations served both to underline their taking place in the international system, and to give further authority to the voice of Britain.

At the same time however, the War had resulted in the Dominions being even less inclined than before to have their foreign policy to be decided in London. Canada was perhaps the most striking example of this, where the Prime Minister from 1921, Mackenzie King, firmly stated that Canadian attitudes in foreign policy were to be shaped by those of the United States rather than those of the United Kingdom.[37] The feeling that they had had to suffer from European power politics in

---

[37] Howard, Michael (1972), ch. 4-5

which they took no part, inspired them to declare their right to stay far outside the conflicts arising out of European power politics.

In terms of land area, the Empire had expanded. What was further, in many of the areas of the Empire the need for British military presence was still great, despite the recent defeat of all her main enemies. Among such areas of potential conflict where large numbers of British troops were needed were not only the occupied Germany and Constantinople, but also areas where the British claims to supremacy were questioned with increasing hostility; India, Palestine and Ireland to name a few.[38]

In sum then, the defence liabilities facing Britain after the First World War were enormous. So were, however, the economic liabilities too. In order to try to get the British economy back on its feet again after the War, the Lloyd George Government decided that both taxes and expenditure had to be cut dramatically, and that came to affect defence expenditure as well. Thus in the years between 1920 and 1922, at a time when the strategic situation called for some kind of military presence in almost all corners of the Empire, annual British expenditure on the Armed Forces was reduced from £604 million to £111 million.[39]

One of the principal factors of world security policy that made this possible was, of course, the fact that if the War had weakened Britain, it had weakened her potential enemies even more. In fact, what made Britain look so strong was the devastation of all the other Great Powers of Europe. Against this background the famous Ten Year Rule of British defence policy was formulated in 1925. It stated that at any point, the standing assumption would be that there would not be any major war within the next ten years from that date. This assumption would rule until the Foreign Office or one of the fighting services took the initiative to change it.

The combination of increased liabilities and shrinking funding for the Armed Forces naturally meant that some scheme of priorities would have to be followed. At the same time the conditions for military defence had changed. During the First World War, the possibilities of air warfare had been demonstrated. A Parliamentary

---

[38] Howard (1972), ch. 4-5
[39] Ibid.

80

committee set up in 1923 to study the need for an Air Force in the Home and Imperial defence concluded: "Air power holds within itself the possibility of bringing about an early termination of a European War."[40] Thus, in the early 1920's, the Royal Air Force, starting from next to nothing, was built up to a considerable strength. The then prevailing technological limitations for air fighting meant that air power in reality was not very effective as actual defence. The defensive use of a strong air force was that of deterrence. Keeping a large British air force with the capability of serious destruction of both military and civilian targets would deter any potential enemies within striking distance from attacking.

For the defence of the Empire however, the strategy was still dependent on above all the Army and Navy. Especially for the defence in the pacific region, the main strategy was relying on the Royal Navy, and its new major base in Singapore. The main concern of imperial defence for the Army was the defence of India, where the main threat planned for was attacks from the Soviet Union on the north borders of India. With the growing nationalist sentiment in India, the standing of the Indian Army became ever more a concern for British strategists, as the British element in it became smaller from 1927 with the process of 'Indianisation' of the Army.[41]

The process of downscaling the Army and Navy and at the same time increasing the role of the Air Force in this way also meant putting a greater emphasis on the defence of the United Kingdom than that of the Empire. It showed a gradual reorientation of British defence policy towards Europe. This was further emphasised by the Locarno Treaties of 1925, by which Britain assumed rather extensive responsibilities in maintaining the European balance of power. In guaranteeing the safety of the German-Belgian and German-French borders from attacks from either side, Britain showed her conviction that in order to defend the Empire, first and foremost the United Kingdom would have to be secured; and that the key threats to British security came from the continent of Europe. In the words of the Chief-in-General of Staff, Sir George Milne, the assumed role of policing peace keeper was

---

[40]Ibid. p. 84
[41]Ibid. p. 92

"...only incidentally a question of French security; essentially it is a matter of British security. ... The true strategic position of Great Britain is on the Rhine."[42]

As the recovery and re-armament of the belligerents of the First World War gradually gathered pace, the stability of the peace again became less convincing. Examples illustrating this is that in 1932, during the great Disarmament Conference in Geneva, Britain decided to cancel the Ten Year Rule. The Italian invasion of Abyssinia in 1935 and the German reoccupation of the Rhineland in 1936, as well as growing tension between Britain and the expansionist Japan further illustrated that the world security order was going back to the pre-war state of balance of power, rather than the dream of collective security. The problem was of course that Britain was in no position to be a powerful counterbalance in this game.

In the years after 1932, when the British rearmament slowly started, up until the outbreak of War there was an uneasy awareness, both from the side of the Services and the politicians, that Britain simply could not afford to be drawn into another major war. The implications of the strategic situation for overall British foreign policy were clear; if at all possible, Britain would have to stay outside a major conflict against Germany, Italy and Japan. At the same time, with reference to the balance of power, she would have to make as much as possible from the links with the powers outside Europe who kept on claiming their unwillingness to once more become involved in a European war, i.e. the Dominions and the United States.

At the same time, as we have seen, the emphasis of British defence policy gradually shifted from a strategy of Imperial defence based on the Navy, to one of primarily defence of the United Kingdom, if necessary on the European continent, based on the Army and Air Force. It was in line with this 'continental commitment' that Britain entered the War in 1939.

Though it can not be seriously argued that all aspects of international relations, including telecommunications policy, were determined by the British defence policy, it is equally far-fetched to assume that it did not have a serious impact on other fields as well. The strategic situation set a framework for international relations, which affected even such 'apolitical' fields as that of international telephony. Furthermore,

---

[42]Quoted in Howard (1972), p. 94

82

the desire to have very close links with the Dominions and the United States both directly and indirectly had some influence on the demand for telecommunications with those countries. It is against that background the 'extra-European outlook' must be seen.

4.7 Summary

In this chapter we have seen how a British set of preferences were formed, regarding the European telephone co-operation within the CCIF. The British telephone system was initially built up in private hands, due to the Post Office's unwillingness to invest in risky new technology. Only after the inter-urban telephone system began to threaten the revenue from the Post Office operated telegraph system did they show any interest to become a telephone operator as well. In a court ruling in 1880 it was decided that a telephone really was another form of telegraph and therefore, legally, was included in the Postmaster General's monopoly rights.

The attitude within the Government and Post Office towards state involvement in telephony was divided between what might be described as 'restrictionists' and 'expansionists'. The Post Office's acquisition of the British telegraph system in the 1860's had been an expensive business. Senior people in the Post Office as well as the Treasury were therefore reluctant to invest new money in a national telephone network. A more positive view of the telephone as a means of communication with potential to take over the importance of the telegraph, was found above all within the technical staff of the Post Office.

Initially, the favour of Government was with the restrictionists, and even after the Post Office's monopoly rights to operation had been established, they were unwilling to invest in the expansion of the telephone network. Instead, the private operators were allowed to continue under a license from the Post Office. As a means of protecting the telegraph revenue, the Post Office retained the right to trunk line, i.e. long distance, operation. Furthermore, an option of purchase of the private networks by the Post Office was included in the license agreements, and after having decided to take over the British national telephone system in 1905, they finally did so in 1912.

The insecurities of the future of the national telephone network acted as a disincentive to invest for the private operators. During the last decade or so of private operation, the telephone network deteriorated, and the spread of the telephone in Britain lagged behind other western states. Following the Post Office's take over investments in the telephone business remained low, due to the First World War and the 1920's depression. This was coupled with a view, primarily from the Treasury, that telephony was a risky investment with unsafe returns. As a consequence the British telephone system compared to those of many other European states suffered from insufficient capacity, especially in terms of trunk line development. Furthermore the restrictionist view of the state's telephone operation put pressure on the telephone branch of the Post Office to bear its own costs. This 'business operation' of the Post Office at times put Britain on a contrary course to that of the other member states in the CCIF, in as far as the British often advocated high rates for international calls while other states regarded their telephone networks as national infrastructural investments, and a means of providing cheap communications to its citizens.

Loosely connected to the insistence on business principles of the British telephone operation was Britain's propagation of free market principles for international lines in Europe. In part this can be attributed to ideological explanations of the free market system as something good in itself. An additional explanation comes from the fact that Britain's location on the fringe of Europe meant that almost no transit traffic passed through Britain. Therefore a system whereby states competed for transit traffic would be a better deal for Britain than the existing European cartel-agreement.

Another situation of conflicting interests within the British government concerned the relation of international telephony to the more general demand for international communications. The technological limitations of telephony during most of the inter-War years confined telephone conversations to the European continent. This was not necessarily the primary interest of the Government in general, who for reasons of matters of economy, security, and the Empire demanded reliable communications with areas outside Europe.

If we thus are to conclude what constituted a 'British position' in the CCIF, we find three general consistencies in British preferences. Firstly there was an unwillingness to invest extensively in the telephone system, coupled with an insistence that the telephone branch of the Post Office should bear its own costs. Secondly Britain in the CCIF wanted to promote a system of competition for international traffic between the European states. Thirdly the British position was characterised by a greater demand for extra-European communications than those of many of the other member states.

5. British Strategies

After having identified a number of major British objectives in their dealing with their European counterparts in the CCIF, we shall now turn our attention to how the British representatives pursued these aims. In the process of influencing the European regulatory order, five major strategies can be identified. These different strategies are in many ways related, but in some parts different objectives called for different strategies. In this chapter I will outline the main features of the respective strategies, and then give examples of how they were carried out.

5.1 Shifting focus away from cable telephony

As mentioned in the previous chapter, the British demand for international communications was to no small extent linked to contacts with the United States and the Empire. Since the technological limitations of cable telephony meant that it effectively was restricted to continental Europe, one British strategy was to devote its energy to other means of communication.

Generally in Europe during the late 19th century, the development of new means of communications, such as the railway and the telegraph, were seen as key resources in building up or maintaining national strength and power. The spread of the telegraph helped transform the economies thoroughly, and were important in the development of new goods and services.[1] In many cases this led to a strong state regulation and control over the telegraph systems.

The development of the British telegraph system in the early 19th century differed, both from the American experience and that of her European neighbours. In the United States the economic and geographic structure of the country, with large economic centres with strong links to each other and at the same time great distances between them, meant a great demand for telegraphy and a rapid growth of the networks. The market structure was rapidly concentrated into a few dominant operators, who could control the market and keep tariffs high enough to stimulate further expansion of the system.

---

[1] Chandler (1990), ch. 3

In continental Europe, the state took a more active role regarding the national telegraph networks. At first the strategic importance given to the telegraph meant that many states were quite unwilling to give their subjects access to the new medium, something which, together with the generally lower demand for communications, slowed down its growth on the continent.[2]

In Britain the government's telegraph policy was originally more *laissez-faire*, with a large number of smaller competitors. The more open, even if indifferent, attitude, resulted in a faster development of the telegraph in Britain, and in 1851 the number of telegrams sent per person in Britain was double that of her closest European rival, but still a long way behind the United States.[3]

Gradually the attitude of the European states became more positive, and in a number of European countries the state took an active part in building up and operating the telegraphs. In Britain this meant that the Crown bought up the domestic telegraph companies, and by the Telegraph Acts of 1863 and 1868 established the Post Office's monopoly rights to operating telegraphs in the United Kingdom.

This not only gave the state direct control over the telegraphs, but also opened for direct interventionism. The great importance given to the maintenance of a good telegraph network induced the government to give direct support in excess of what could be raised from normal revenues. Since the Post Office came to be the operator of the telegraph as well, a preferred instrument for such support was a substantial cross-subsidy of the telegraphs from the postal operation.

The telegraph was an important communication link to all parts of the Empire and Dominions, and as such it also had a serious strategic value. This is illustrated by the British encouragement to construct telegraph lines all over the Empire. For instance in the case of the Red Sea and India Telegraph Company, the encouragement went so far as to include an annual subsidy.[4] This enthusiasm over the telegraph as a means of extending Imperial communications grew more marked with the invention of the submarine telegraph cable, which made inter-continental

---

[2] Foreman-Peck and Millward (1994), pp. 53-61
[3] Foreman-Peck and Millward (1994), p. 55
[4] Foreman-Peck and Millward (1994), p. 144

communications possible. In fact, an important share of the world's system of international submarine cables were British owned, something Germany learned the hard way in 1914 as their telegraph communications were cut.[5]

Foreman-Peck and Millward (1994) test whether the subsidies to the telegraph system were justified, i.e. whether they gave an increased telegraph output, and conclude that they did not. Thus the justification for those subsidies was not economic.[6] What is important here, however, is rather the government's perception of the telegraph as a crucial resource, and their willingness to invest in it.

The crucial factor in Britain's emphasis on alternative technologies to cable telephony was of course the limits of its geographical reach. As we have seen, the thermionic valve and telephone repeaters had made distance less of an issue in telephony. But still construction of submarine cables over long distances brought severe problems. Cable construction was an expensive business, and in the case of very long submarine cables, the technology simply was not developed. This meant a constraint to the European continent, which was not to be broken until after the second World War.

What potentially could free telephony from those restraints was radio technology. Britain's demand for extra-European communication meant a willingness to support attempts to develop radiotelephony. Already in the last years of the 19th century, Post Office engineers had experimented with using water as a conductor without much success. When Marconi arrived in Britain in 1896, his radio experiments and refining of radio technology were sponsored by the state, as both the Post Office and the Royal Navy saw the potential in using the ether for transmission. The radio telephone proved to be an efficient means of rapid communication, and both the Royal Navy and the RAF soon acquired wireless telephones to prevent collisions.[7]

The use of wireless as an alternative to cable telephony was also soon realised. In 1915 an experimental message was sent successfully from the Eiffel tower to

---

[5] Headrick (1981), ch. 11
[6] Foreman-Peck and Millward (1994), pp. 144-147, 157-161
[7] Foreman-Peck (1992), p. 171

88

Arlington, Virginia. As the new technology grew more successful, the Post Office once more started to protect its monopoly rights to electric communications, and in a move similar to when the cable telephone business was monopolised in 1880, the Post Office withdrew Marconi's license to operate.[8]

In 1927, the first radiotelephone service between London and New York was opened by the Post Office, and gradually was extended to all of Great Britain and the United States, followed by Cuba and Canada. The year after, the transatlantic service was extended to Mexico, followed by Australia in 1930.

This new feature of international telephony put Britain in telephone contact with the largest part of the Empire, and in practice provided a link independent of the European cable network. The European system was however being connected to the wireless telephone system, so that subscribers all over Europe effectively could be put in contact with distant parts of the world.[9]

In summary, the greater British demand for communications with the Commonwealth and Empire and other areas outside Europe showed in a relatively greater interest in the telegraph and radiotelephony. Once radio telephone links had been established with the rest of the world, this meant that it suddenly had become possible to be in telephonic contact with all parts of the world. Such was the success of wireless, that after the opening of the London-New York link, 65% of the submarine traffic was lost to the new technology.[10] This illustrates the demand for extra-European communications. Further it meant that Britain's dependence on the cable network, and therefore her dependence on her European partners, lessened dramatically.

Still, however, a number of problems lingered with the technology of radiotelephony. Limitations of the spectrum meant that it could not be substituted for all cable connections, but was used only where cable telephony was impracticable, such as in long distances over water. Further the sound quality of radio was both

---

[8] Foreman-Peck (1992), p. 171

[9] Britain was in the 1930's becoming a major node in the transatlantic telephone system, which made a contemporary Post Office document revising the development of the overseas telephone contacts boast of London being "...the switching centre of the world.", Post Office (1931)

[10] Foreman-Peck (1992), p. 172

inferior and unreliable. In the links where cables could be used, that was still a better technological choice.

5.2 Shifting focus away from Europe

The more international, as opposed to European, outlook also showed in the close ties with above all the American telephone system. This naturally implied close links also to the Bell-system, with International Western Electric as its European branch of its appliances industry. The Institution of Electrical Engineers was a British forum for spreading the gospel of electrical engineering. The American network, being by far the most advanced with international standards, held a lot of prestige internationally, and the representatives of the Bell system were very active in the Institution. This clearly influenced the British engineers, and helped in strengthening the links to the British system.

In trying to realise their goals, the British telephone administration gradually tried to shift the focus of the very Eurocentric CCIF, and involve the United States in the international co-operation.

At the 1938 International Telegraph Conference in Cairo, a rare example of written instructions from the Foreign Office to the British delegation noted that the United States had not signed the International Telegraph Regulations, but had hinted that they would be able to do so if the regulations were divided in two parts; one more general provisions and the other regarding more detailed workings for the telegraph service. In that case the US could sign the first part. The instructions were very clear in saying that any scheme which the United States will put forward for this purpose shall be considered sympathetically. The instructions further stated that the delegation should keep in close touch with the Dominion delegations. Regular meetings should be arranged; and efforts be made to secure friendly co-operation.[11]

5.2.1 Report charges

Another example of this strategy is provided by the example of reaching a common standard for so-called 'report charges'. When setting up radiotelephone calls, the

---

[11]*Instructions to the United Kingdom delegation to the Telegraph and Radiocommunication Conferences of Cairo, 1938, pp. 11-14.* PRO HO 257/99.

90

procedure typically would be that the call was first ordered by a subscriber, who then would be called back when the preparations for the call were made. An example illustrates the procedure[12]:

Personal call Amsterdam 12345 (Mr, Stroerr) to New York Stuyvesant 5678 (Mr. Van Riebek).

I	Mr Stroerr orders the call from his Amsterdam station
II	Amsterdam passes call to London
III	London passes the information to New York
IV	New York calls Stuyvesant 5678, asks for Mr Van Riebek and obtains report that he is ready to talk
V	New York advises London that Mr Van Riebek is ready to talk
VI	London passes report to Amsterdam exchange
VII	Amsterdam exchange rings 12345 in order to pass report that Mr Van Riebek is ready to talk, and the call is set up.

The setting up of the call thus included a certain amount of preparation, and required holding the line open before the call could actually be put through. The transit countries would of course have to be compensated for this in some way, but in the cases where the call was cancelled for some reason, it was not obvious who should pay for this 'report charge'. Different countries used different rules for at which stage a call could be cancelled without the person ordering it being charged for it, which caused confusion.

In July 1933, the German administration took the initiative to bring the varying practices in line. In a letter to the British, French, Italian and Dutch Administrations, and to the Spanish Telephone Company, a new set of uniform rules was proposed.[13] The CCIF had already issued recommendations on uniform rules relating to the European service. In the matter of inter-continental calls over radiotelephone links, however, these rules were not applied. The German proposal was to bring the radiotelephone service into line with the European system.

---

[12]Example taken from: "Memorandum detailing certain phases of British Radiotelephone Service Procedure", 17 July 1936, *BT Archives, Post 33/4704*

[13]Letter from the Reichspostminister to the Secretary, General Post Office, 11 July, 1933, *BT Archives, Post 33/4704*

The French and Belgian administrations soon expressed agreement with the German proposal. The Post Office, however, chose to take a different view. In a letter to R. W. King of AT&T, it stresses that the priority for the Post Office is to conform its practices to that of AT&T rather than that of the other European Administrations:

> "...we are making slight amendments to our own procedure which will make it approximately the same as yours...We shall next suggest to our partners in the Dominion services that they might like to adopt the same procedure, and I do not anticipate any trouble in this respect. We should then be in a position to advise Germany and the other countries of the procedure utilised by your Company, ourselves and our other partners..."[14]

The reply from the AT&T Overseas services states that the German proposal "...is undesirable because it is entirely out of line with our land line practices and practices on our existing overseas services."[15]

The example of 'report charges' shows the policy of the Post Office to adjust to the American system, rather than to the European, and its willingness to focus on extra-European territories in shaping a world standard rather than a European one. In a number of letters to the Dominion Services, the Post Office called for an international standardisation of the report charge procedure, and proposes the common British-American position, without mentioning the European standardisation measures.[16]

This is typical of the British strategy in trying to shift the focus of international telephone co-operation to a world setting. In a large number of cases of standard-setting, the Post Office propagated the choice of standards which conformed to the more advanced American system. In a number of instances when their position became isolated in relation to their European partners, they first tried to bring in the Dominions and the Americans into the CCIF, and if that failed, to bypass the European co-operation.

---

[14]Letter from H.G.G. Welch, Post Office, to Dr R.W. King, AT&T Co., 20 November 1933, *BT Archives, Post 33/4704*

[15]Memorandum from the AT&T Overseas Services, dated New York, January 3, 1934, *BT Archives, Post 33/4704*

[16]Letters from the Post Office to the Administrations in Melbourne, Cairo, New Delhi, Pretoria, Canada, and Buenos Aires, dated 21 March, 1934, *BT Archives, Post 33/4704*

5.3 Trying to introduce competition

As the international services in Europe developed in the 1920's, one of the chief problems was, as mentioned before, the shortage of trunk lines. In many of the European countries where telephony still was largely undeveloped, the construction of special lines for international telephony was not always regarded as a high priority. This of course meant that also the more advanced telephone nations suffered from the poor state of the European trunk-line system. If two relatively developed telephone states, such as e.g. Great Britain and Sweden, wanted to establish a direct circuit, they were dependent not only on their own lines, but also on those of the transit countries in between.

What was needed for the construction of a developed European international network was the provision of:

●Adequate financial inducement to the transit country to lay the necessary cables; and

●Adequate motives to induce the transit country to maintain the transit circuits (which its own nationals do not use themselves) with a high degree of efficiency.

In the countries which already had reasonably developed domestic telephone systems, the problem of assigning circuits as special "through" circuits for transit of international calls simply was one of agreeing on a rate which would give the transit country a reasonable rate of return on their investment. In some cases these countries would not even have to lay special cables for the through circuits, but could simply provide circuits out of their reserves on a message basis.

Various methods for agreeing on international rates were used. The one which came to be called 'the CCI system' meant that the rate was simply calculated as the sum of the terminal rates of the countries where the call originated and terminated respectively, and transit rates for the countries through which the call had to pass.

In the case of the less advanced systems however, a new international circuit passing through their territory would mean that they had to lay special transit circuits. If the traffic on the new circuit developed slowly, they then ran a risk of having invested in a cable that cost them much-needed investment capital but was not used of their own nationals.

One way around this under the CCI-system was for the terminal telephone Administrations to guarantee a minimum number of calls through the new circuit. In this way a certain amount of the risk could be shifted towards the countries who desired the cable in the first place. But unless they could get guarantees for all of the circuits, or they could count on using the rest of the circuits themselves, they still ran a risk. In these cases it would be more beneficial for them to lease the circuits to the terminal authorities for a fixed rental.

For the terminal authorities, a fixed rental of cables meant accepting the whole risk of the investment which, nonetheless, benefitted the transit countries which could use the spare capacity in the cables. One further problem on behalf of the terminal countries was the problem of maintenance of the circuits. Of course some form of penalty could be agreed upon for frequent or lengthy breakdowns of the services, but it was more difficult to regulate less obvious cases of poor maintenance, which nevertheless decreased the value of the service and could mean loss of traffic.

In the case where a call had to be routed through several transit countries, the system with leased cables also could mean that if the service broke down due to poor maintenance in one of these, the leased cables through the other transit countries would become useless to the terminal authorities. The CCI system on the other hand of course meant that there was a strong incentive for the transit countries to keep a good maintenance on their through circuits, especially when there were no minimum traffic guarantees given.

In agreeing on the conditions for transit traffic, the European co-operation through the CCI developed rapidly during the 1920's, so that these procedures became standardised. What could have been a daunting task; to reach an agreement in multilateral negotiations, ran fairly smoothly due to the standardised manner in which the agreements were made.

An example of this is during the setting up of the Anglo-Swedish circuit in 1926, when the representatives of the British, Dutch, German and Swedish authorities could reach a principal agreement within an hour.[17]

---

[17] Post Office (1926), Memo on the leasing of transit circuits. BT Archives Post

5.3.1 Rate conflict

The Anglo-Swedish circuit is perhaps not such a good example of the negotiations over international connections running smoothly, though. Despite the fact that the principles on which the new circuit should be operated were quickly and readily agreed upon, the fixing of the rates became a problem, and in fact amounted to a fairly serious conflict in the European co-operation.

The question of which principles should be used in the calculation of rates had before been the matter of discussion and some disagreement at the CCI conferences, but in the case of the Anglo-Swedish circuit, these disagreements for the first time meant that there became something of an open conflict, hindering the development of international telephony.

In this conflict the opposing views were held by, on the one hand Britain, and on the other, the Continental administrations, led by Germany. The Continental view was that the total rate between two countries should simply be the sum of the terminal and transit fees of the administrations concerned, and that these fees should be laid down by the respective administrations as 'avis' with the CCI. Furthermore, the Continental administrations pressed for these rates being calculated uniformly for all administrations at a fixed price per 100 kilometres of crowfly distance between the points of entry and exit for the cables of each circuit in that country, plus a fixed terminal exchange charge at each end.

The view of the Post Office was that the rate for any service would have to be fixed according to demand. By their view it was not only reasonable but also desirable that the rates should be based on the user's willingness to pay, rather than the cost of operation. The only other administration to support Britain in this view was the Swedish Telegrafverket, but even they disagreed with the rates policy of the Post Office.

The principal complaint of the Continental administrations was that the British terminal charges were much too high, and so slowed down the widespread use of the international service. In 1926, the German dissatisfaction with the British terminal rates led them to such a drastic measure as to price discriminate against Britain.

The agreed upon and almost universally adopted rate of the Continental administrations was .60 gold francs per 100 kilometres for transit traffic. As an attempt to put pressure on the Post Office to lower their rates, the Germans started charging 1 gold franc per 100 kilometres for both transitory and terminal traffic originating in Britain. Germany tried to persuade other administrations to do the same, and were soon followed by Belgium.

In the following months the tone sharpened between the German and British administrations, and the result was that though a principal agreements on the opening of the Anglo-Swedish service, they failed to reach an agreement on the rates, and the opening of the service had to be cancelled.

In a letter to the Comptroller and Accountant General of the Post Office, Sir Henry Bunbury, the Director General Arendt of the German Reichspost complained about the British position to charge lower unit charges with increasing distances of the call. This ran against the cost-based pricing principle, since the unit cost of long-distance calls actually was higher, since they demanded higher quality cables etc. This would then require

> "...that the charges for the traffic over short distances shall be raised higher above their actual cost. This tariff policy is easily carried out when only one Administration is concerned in the service, as for example, in America, the American Telephone and Telegraph Company. It appears to me, however, to lead to great difficulties as soon as several Administrations must co-operate as in the European services."[18]

Considering that the various Administrations primarily would see to the interest of their own nationals, it would be untenable for them to charge a higher unit charge for the terminal traffic originating in their own country or, for that matter, domestic calls, than for the transit traffic that only passed through their territory. Mr Arendt further pointed out that not only would it be unfair to levy different charges for the same service, it was also forbidden according to article 71 K, paragraph 4 of the Paris *reglement*.[19] This was probably not an argument the British Administration was too open to, considering that the Germans were actively persuading other Administrations to do just that against the British.

---

[18] Arendt (1927), Letter from Arendt to GPO, dated Berlin, 21 April, 1927, BT Archives, Post
[19] Ibid.

96

The different opinions on this subject between Britain and Germany should naturally be related to the geographical locations of these countries. Until the Anglo-Irish service opened, Britain had no transit traffic at all, except the radiotelephone services. If a system whereby European countries could be made to compete for transit traffic was introduced, Britain of course would benefit from the lower tariffs, while at the same time keeping the terminal fees on a high level to cross-subsidise the domestic operation.

This, naturally, was not in the interest of the Germans. By its geographical location and its well developed telephone system, Germany was probably the most important transit country on the continent. This not only made the German interest in fixing the transit charges favourably, but also meant that the German Administration had an important position as a powerful player in almost all continental telephone affairs.

The British position was, as we have seen, one of more reliance upon the market and demand to decide the rates. In a way it could be said that this cognitive notion about how the telephone system could be run most efficiently played an equally important role to that of the geographical location of Britain as a country with important terminal but almost no transit traffic in the formation of British positions on the issue of rates.

It is clarifying to bear those two background motifs in mind when looking at the British attitude towards the German position of power within the European telephone system. From that outlook it was perceived as generally bad for the market to have one actor on the European arena with such a powerful position, and the British strategy towards addressing this was of trying to introduce some measure of competition into the thoroughly regulated European system.

In addition to these explanations to the continuing clashes of interest between the British Post Office and the German ReichsPostamt, one must also account for the higher level politics of foreign and defence policy.

5.4 Influencing through expertise

Most of the work within the CCIF was and still is of a rather specialised, technical nature. Issues that need to be studied in detail before the Plenipotentiary conferences could decide upon them were often remitted to expert study groups of Rapporteurs, and by supplying expert technicians to these study groups, the member states could influence the CCIF decisions before they were on the general agenda.

Worth noting here is that not even these extremely specialist technical matters can be dismissed as completely apolitical. An example of a conflict within such a commission of Rapporteurs regarding the choice of a common transmission unit can provide an insight into the political entanglements of such issues.

5.4.1 The choice of transmission unit

In 1924, the CCIF started to address the issue that there was no common practice for measuring transmission over telephone lines, and accordingly they set up a study group to deal with the problem and give recommendations to the next conference. Designated Rapporteurs of the group were Dr Breisig of Germany and Colonel Thomas Purves, Engineer-in-Chief of the Post Office.

As their work turned out there were two candidates for the common unit. One was the so-called natural unit, employed in applied physics, β. The other was a new unit, suitably called TU, transmission unit. The β unit was the one previously used by telephone engineers. The meaning of it is a measure of current or voltage ratios. β is the loss in percentage of current or voltage from the beginning to the end of a section of unit length. The TU was introduced by the American Bell System in the early 1920's. The advantages with it was that it is a measure of power, rather than current or voltage ratios, and since the fundamental requirement of a telephone system is the transmission of power, that seemed appropriate. Thus comparisons between different types of lines could be made directly by using the TU, without need to consider other factors. It does also have the advantage of making possible comparisons of large power ratios without advanced mathematical and technical knowledge. Lastly, since the system had been adopted by the Bell System, the advantage of Europe adopting it as well would be to create a world-wide standard.

These points taken together were the arguments Col. Purves pleaded for the adoption of the TU. The German Rapporteur however, was not so easily convinced. Dr Breisig's arguments against the adoption of the TU was that it would deviate from the practice in other departments of applied physics. In his report on the matter to the designated study-group, his conclusion was that "...more than a certain amount of convenience for people not engaged in theoretical and experimental research ought to be demonstrated in order to justify such an exceptional departure from general methods".[20]

Since no unanimous agreement could be made within the group on the choice of unit, the committee resorted to the quite unusual means of voting. The outcome of the vote was:

•For adopting the natural, absolute unit, β; Germany, Austria, France, Italy, Sweden. Switzerland, Czechoslovakia, and Yugoslavia.

•For adopting the transmission unit, T U; Great Britain

•For using both β and T U; Belgium.[21]

In the absence of a general agreement, the matter had to rest, and no common unit was adopted. That result seems almost incomprehensible. The issue of finding a common unit was quite pressing, since it was very inconvenient for the European Administrations not to have a common practice for measuring transmission. All international lines had to be tested, in order to find out whether they were good enough for using as long-distance connections. The question of which unit was chosen obviously was less important than the object of actually agreeing on one, single common unit (though apparently not so to the Belgian Administration). Yet the representatives could not make an agreement.

The answer lies once more in Britain's unwillingness to bind herself too closely to the continental European system. If the Americans used a new unit which both was better designed for telephony and more user friendly; and the adoption of that unit would as it were create a world standard rather than a European one, they saw no reason to opt for the β unit. The other European Administrations on the other hand, saw no reason to abandon the unit many of them already used. The fact that the TU was widely used in America had little real importance to them, since there was no

---

[20]Comité Consultatif International des Communications Téléphoniques à Grande Distance, Permanent Commission, Second Conference (held in Paris, November 24th-December 1st, 1924), Minutes of proceedings, English version, *PRO Kew, HO 257/51*

[21] Ibid.

telephone contact with America anyway. Their standpoint can be explained as marking out their identity, both as a common European system, and, as we have seen from Dr Breisig's argument, electrical engineering and telephony as science, rather than business.

The example of the choice of transmission unit shows how Britain tried to influence the CCIF's decisions through their expertise. In reality many of the important decisions were made in the expert study groups. It was there that the concrete suggestions which were to be decided upon by the Plenary Assembly were formulated. By influencing the formulation of recommendations from the study groups, member states could influence the agenda of the CCIF.

The British Post Office employed technicians of high standing. Through the Institution of Electrical Engineers, they also had close contact with American engineers from the Bell system. This gave them the opportunity to have representatives on important study groups, in many cases as Rapporteurs. At the same time, supplying expertise to the CCIF's study groups represented a cost. The fact that British engineers were represented in many of the commissions indicates that the Post Office regarded it as a good investment, and a workable strategy for influencing the European telephone network.

5.5 Blocking decisions of the CCIF

The low-cost objective meant that Britain was unwilling to accept European standards that also required heavy investment in the national networks. It also meant a reluctance towards spending money on the international administrative apparatus, such as the overhead costs of organising the European co-operation through the CCIF.

An example of this is found at the 1938 International Telegraph Conference. The Telecommunication Convention then made provision for three Consultative Committees, namely Telephone (known as the CCIF), Telegraph (CCIT.) and Radio (CCIR.) Their functions as laid down in the Regulations differ. The Telephone Committee was established for the practical purpose of organising the European telephone service; and it is authorised to study technical, traffic and tariff questions. The other two Committees were restricted, the Telegraph Committee to the study of

100

technical questions and operating procedure and the Radio Committee to the study of technical questions only. Italy then proposed to make the functions of the three Committees uniform by empowering the Telegraph and Radio Committees to discuss tariff and traffic questions. The instructions to the British state that: "These extensions of functions are not considered desirable. The two Committees have done valuable work on technical and allied questions and have always had a large programme of work....It would... largely increase the number of persons attending the meetings of the Committees with consequent direct and indirect expense to the various Administrations."[22]

The strategy of blocking decisions was a sort of last alternative, something to use if the efforts at controlling the earlier stages of the decision making process had failed. In a way it is paradoxical that the strategy worked at all. In the statutes of the CCIF there is no rule stating that all the decisions of the Committee had to be taken unanimously.

Further, and this might be worth repeating, all the decisions of the Committee were recommendations, i.e. the member states were free to ignore the rules if they wanted to. That could be expected to work as a safety valve, or something which should enable the CCIF to make decisions even if one member state was against it. Still Britain managed to veto decisions.

It is difficult to precisely pinpoint why or how they could do that. The answer seems to lie in the organisational culture. One writer states that "...a tacit unanimity rule exists in all the voting fora of the ITU."[23] In practice, the prime objective of the CCIF was to find common solutions to problems. It is possible to envisage a European telephone system where one country decided to stand outside the co-operation, but that would mean complications and difficulties to all the others as well. That, taken together with the organisational culture, where the participants in the organisation saw their justification in the finding of common solutions, apparently was powerful enough for letting one country block its decisions. That strategy however was a weapon to be used sparsely. Supposedly the consensus

---

[22]*Instructions to the United Kingdom delegation to the Telegraph and Radiocommunication Conferences of Cairo, 1938, pp. 10-11.* PRO HO 257/99.

culture would not have been strong enough to let the same country repeatedly hinder agreements in different areas; but occasionally and as a last resort, the strategy worked.

5.6 Summary

In this chapter we have seen five different strategies of British action within the CCIF outlined. They are in many ways related, but to some extent the different objectives presented in chapter 4 required different approaches. The first of these was to shift focus away from cable telephony and towards other means of communication. The demand for extra-European communications, which could not be fulfilled by cable telephony during the inter-War period, meant that British communications policy included devoting attention to the development of alternative communications.

The telegraph with its lesser demand for bandwidth and clarity of signal transmission was a technology where submarine cables had been developed to a state where intercontinental communication was possible. This had led to the establishment of a world-wide net of telegraph cables, which to a substantial degree was British owned, and in some cases subsidised by the British state. In the same way, Britain encouraged the development of radio technology, and when radiotelephony became a working alternative to cable telephony, Britain was quick in establishing links to the rest of the world.

Another strategy associated to this one was to internationalise the telephone co-operation in the CCIF away from its highly Eurocentric origins. The demand for communications with the rest of the world is of course an aspect here. That demand made it important for Britain to reach truly international standards and interconnection, rather than achieving co-ordination on a European scale. One further aspect of that is that inclusion of other parts of the world into the game of international regulations would give Britain a relatively stronger position visavi the other European states. Even if the relations with the Dominions often were rather tense during this period, they could at least be counted on to have more in common

---

[23] Savage (1989), p. 12

with Britain than with other nations in Europe. The same sort of calculus would also hold true for the Anglo-American relationship.

That special relationship between Britain and the United States was thus played upon heavily in the negotiations of the CCIF. The American telephone system was by far the most advanced in the world, and in the British Institution of Electrical Engineers (IEE) American engineers, more often than not from the Bell system and AT&T, enjoyed high prestige. This clearly influenced the British engineers who were directly involved in the CCIF negotiations. In a long series of cases when the others in CCIF tried to introduce separate European standards, the British representatives opposed this, and instead argued for adopting AT&T's standards and procedures in Europe as well.

The third strategy was to try to introduce competition for international traffic between the European states. The standard procedure for deciding rates for international calls was to calculate a uniform price per 100 kilometres of transit traffic plus a terminal charge at each end of the line. Britain instead propagated a more flexible pricing, based on demand. Her main opponent in this was Germany, and the disagreements over rate policies in the late 1920's led to price discrimination between them and a standstill in the opening of new lines in Europe. Explanations for this serious clash of interests between Britain and Germany include the fact that whereas Britain had almost no transit traffic at all, Germany was the most important transit country on the Continent, but also ideological or cultural differences as well as matters of foreign and defence policy played their part.

A fourth strategy was to influence through expertise. The technical work in the CCIF was carried out by expert study groups who closely examined specific questions of co-ordination, and then put forward suggestions to the General Assembly to decide upon. This clearly made these study groups important arenas for influencing the agenda of the CCIF. Through having plenty of highly skilled electrical engineers and a good relation to AT&T, Britain possessed a certain expertise in most telephonic matters. This gave them the opportunity to supply experts to the study groups and thus influence the CCIF's work.

In cases where all other strategies failed, the British could turn to a last attempt to have a say in the co-ordination: blocking the decisions of the CCIF. Perhaps the most surprising thing about that strategy is that it worked at all. Formally there was no unanimity rule in the CCIF. Nevertheless there was a strong culture of consensus among its members. The need for co-ordinating all the parts of the network in order for the whole to function satisfactory was also an incentive for the CCIF to reach agreements which all its members could share, but it is probable that a country which consistently had obstructed the decisions of the organization sooner or later would have been ignored. This was thus a strategy to be used sparsely, and only as a last resort.

6. Conclusions

Against the background of the preceding chapters about British telephone policy during the inter-War years, I would in this chapter like to summarise my findings and, if possible, try to draw some more general conclusions about the actions of states in international organisations. The first section of this chapter deals with how the national interests were formed from a large number of diverging interests on the national level. Secondly I turn to the issue of how Britain acted to put forward those interests on the international level.

6.1 Formation of objectives

A closer examination of how Britain's objectives in the CCIF were formed reveals that the state is not suitable as the primary unit of analysis, as is often the case in realist or neo-realist studies. The process whereby national positions are reached is far from the clear-cut picture of a nation's struggle for economic, military or technological supremacy. Instead a number of different actors, each with their own goals, were tied to each other in a set of interdependent relations.

If we try to adopt Putnam's concept of a two-level game (as described in section 1.2) the national level in the case of Britain's policy on international telephony could be expressed as in figure 6.1

Figure 6.1 A model of the national level

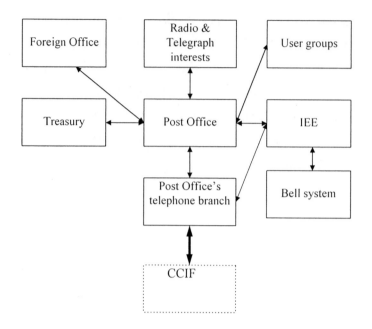

The institutional set-up led the actors into what has been described as antagonistic co-operation. The British Government had certain ends which it wanted to achieve in international relations generally, but lacked the necessary expertise to deal with the technicalities of telephone regulation. The telephone branch of the Post Office in their turn also had certain goals with the CCIF co-operation, but were of course at the same time dependent on other domestic actors. Throughout the inter-War era we see an ongoing struggle; both within the Post Office for a more detached position for telephone and telegraph operations visavi the operation of the postal system; and at another level a struggle of the Post Office for increased independence from above all the Foreign Office and Treasury.

The process of goal formulation also included other actors, both domestic and foreign. Since the individual actors in many cases not were representatives of only one body, say, the Post Office but also members of for instance the Institution of Electrical Engineers, it is plausible to assume that their preferences also were affected by such influences. The personal networks of which the key actors in the British policy formulation can therefore be thought of as important too in shaping the

106

final outcome. Through the networks of the IEE, the American Bell system also had a link to the Post Office's telephone branch, and through that to the CCIF. Among the other interests who tried to influence the British position can be mentioned user groups, represented for instance by the Chambers of Commerce in the case of pricing policies. Also radio and telegraph interests should be included in the analysis. Both operators and manufacturers of course had an interest in Britain's use of alternative communications technologies.

In the game of negotiating the British position this had a further influence on the British objectives, so that the formulation of British goals was successively shaped and reshaped into a form that reflected some of the influence of all the different actors who at various stages had been in a position to affect them.

6.2 Strategic action

After having identified how the British objectives for the international work in the CCIF was formed, we then turn our attention to how the British representatives tried to gain support within the organisation for their goals. In chapter 5 we identified five strategies which Britain pursued in negotiations over different issues. The strategy of shifting focus away from cable telephony, was related to shifting the focus away from Europe. Limitations of technology, most of all the construction of submarine telephone cables, confined Britain's cable telephone contacts almost exclusively to the European continent. Thus developing and promoting alternative means of communication could reinforce that strategy. At the same time Britain had important interests in international telegraphy and radio. As mentioned in chapter 5.1, international telegraph companies were British owned and even sponsored by the British Government, and Marconi's development of radio technology was supported by the state.

The strategy of shifting focus away from Europe filled two functions. Firstly it reflected the British demand for extra-European communications, above all with the Empire, Dominions, and the United States. Secondly it could serve as a way of increasing Britain's influence over the international regulatory order by bringing states into the co-operation who could be expected to have a more sympathetic view of British interests. This is for instance illustrated by the Foreign Office's instructions to the British delegation at the Cairo conference in 1938 to keep in close touch with the delegations of the Dominions, and to consider American suggestions regarding the Telegraph Regulations sympathetically.

The third strategy was aimed at introducing competition for international traffic. This can in part be seen as an attempt to lower costs for British users at the expense of those in countries where transit traffic constituted a larger proportion of the total international traffic. The promotion of a system of rates based on demand can also be seen as an attempt to satisfy those who advocated the American system of high price and high quality as a means of spreading telephony to business users.

As number four on the list of British strategies we find a more general approach to how Britain tried to influence the CCIF. Through supplying highly skilled engineers to the CCIF's expert study groups, Britain could have an early say in the technical issues as they were discussed before they appeared at the General Assemblies. Since the national telephone administrations did not receive any financial compensation for the work carried out by their experts in these study groups, it represented a real cost to them. That the Post Office nevertheless chose to send experts to a large number of such committees indicates that this was seen as something in the British interest to do.

The last strategy on our list is simply blocking decisions within the CCIF. As discussed in chapter 5.5, it is rather surprising that this strategy worked at all. It would probably not have worked if frequently used, but in some cases Britain turned to this strategy of non-decision.

Hypothetically we can identify a couple of features which would, *ceteris paribus*, make a country more influential in the CCIF negotiations. The consistent logic of Britain's strategic action within the organisation was to use strategies which minimised the importance of such features where Britain was lacking, and made the most of those where she had an international advantage. In addition to playing on those external factors, countries had to rely on coalition building within the CCIF, to gain support for their ideas.

One key feature was a strategic geographic location. The physical nature of the telephone cable network meant that a country which was centrally located in the

European network would have a substantial volume of transit traffic. This also implied that the marginal benefits of having such a country included in the co-operation were higher than those of including a more peripheral country. The difficulties and the costs associated with bypassing a centrally located country implied that it would have a strong position from which to bargain.

In the telephone system of inter-war Europe, the best example of such a country is Germany. Britain's position off continental Europe and with almost no transit traffic at all meant that she had to rely on other sources for influence. The geographic peripherality of Britain can however perhaps help explain some of the British strategies. The development of other methods of communication than cable telephony made geographic centrality less of an issue. With the opening of the London-New York radio telephone link in 1927, Britain's centrality in the telephone system increased, and, as we saw, British officials proudly remarked that London was 'the switching centre of the world'.

The British strategy of shifting her focus of telecommunications away from Europe should also be seen in this light. The underlying logic of that strategy was of course a greater British demand for extra-European communications, but at the same time the creation of an international telephone network which included other parts of the world reduce the disadvantage of being in the European periphery.

Another feature which, all other things equal, can be thought to have given member states in the CCIF a powerful position was a strong domestic technological base. Such technological leadership could give influence in a number of ways. An advanced national telephone network which worked well could serve as a model for other countries who were building up their long-distance networks. That would then of course also mean an advantage in the choice of common standards. Secondly one could imagine that a technological leadership would render a country export opportunities, and thus influence other countries' national standards. Thirdly, by supplying advanced engineers to the expert committees of the CCIF, a country could gain influence over its proposals to the Plenary Assembly.

As we have seen, Britain's domestic telephone network was not very advanced by European standards. Furthermore, the national telephone industries were often seen as national assets in technological development, and were therefore in many cases protected to various degrees. It was precisely that system of independent industries which led to the evolution of national styles in the national networks. In so far as there was any conforming influence from technology exports, the strong actor there was the American Bell system who, through their appliances manufacturer International Western Electric was something of an international market leader.

What Britain did have though, was technological know-how and skilled engineers. The close links with above all the American system, both through direct contacts between the Post Office and AT&T, and through the Institution of Electrical Engineers, meant that British engineers in many cases were familiar with the most advanced system, the American. As we have seen, Britain used this asset to influence the CCIF by supplying experts to its study commissions, and this proved to be a useful strategy.

Apart from these strategies which built on external factors, the various member states had to gain support for their preferences by building coalitions with each other. One way of trying to achieve support is illustrated by the British goal of introducing some measure of competition for transit traffic. In that case the British delegates presented such competition as something universally good, which would benefit the whole of the CCIF and its member states. As it were, the European market for telephony was not very fertile ground for such free-market thinking. As demonstrated by the liberalisation efforts by the EU regarding the European telecommunications network, such a market still (in February 1998) has not materialised; more than ten years after the Commission's decision to make it a high priority goal.

It is worth noting though that the British free-market policy should not be merely regarded as empty phrases to conceal self-interested rent seeking behaviour. As stated before, the British belief in market solutions and independent suppliers was firmly rooted in the official British mind, and most likely was a genuine idea of how things could be made better; in the same way that the Germans generally believed in stronger political control and standardisation.

To sum up the discussion here, I have found that the co-ordination of international telephony in Europe was a political process, where the concerned states had national interests which they were pursuing. The formation of such interests involved a large number of actors at different political levels, and resulted in a set of different objectives. These objectives are not necessarily those which could be assumed from traditional realist assumptions about self interested states. In their efforts to influence the whole of the common European regulatory order, British representatives used a number of different strategies, in part decided by the objectives they were pursuing, and in part by the relative national advantages and disadvantages.

One general conclusion to draw is that the previous view of telephone regulations as something purely technical, and therefore apolitical and uncontroversial, is wrong. Member states did regard international telephony as a field of international politics, and cared about the outcome of negotiations. They had different national interests which they were pursuing, using a set of different strategies to put forward their interests. In this respect this study of the CCIF conforms with the traditional intergovernmental approach to international organisation.

Against this background I conclude that one of the more general results of this study is that fruitful analysis of international co-operation and international organisation requires an in-depth study of the domestic and international setting in which the national objectives are formed. Simplistic assumptions about the state as a maximiser of power, be it technologic, economic or military, will leave important parts of the international agreements outside the analysis, and so to speak define away some of the factors which should be analysed.

6.3 Further research issues

One aspect of the strategic action within the CCIF which seems highly important is the coalition building and mutual agreements between member states within the organisation. Factors like the formation of personal and social networks, as well as bilateral relations beside the CCIF could be expected to be important in this respect. Therefore one area of further research which could shed some more light on how states acted in order to gain influence in the CCIF, would be a systematic treatment of those factors, in order to show if there were persistent 'blocs' within the organisation and if so, how these were built up and maintained. Such a study would give us further insights into the workings of international specialist organisations, and possibly shed some new light on how international politics work.

Another interesting aspect is that of the CCIF as a corporate actor. As noted in chapter 3 the CCIF once established firmly, gradually incorporated new areas on its agenda. Corporate actor theory suggests that when states pool parts of their decision-making powers into an organization, this body then gradually starts acquiring preferences and strategies of its own. It is also reasonable to think that the national representatives who came to form personal networks through the work in the CCIF discovered overlapping interests between themselves, which were not necessarily identical to their national positions. As we have seen, those national positions were reached through complex networks of bargaining between different interests on the national level. A study of how such preferences and strategies of the CCIF were formed and used would provide an interesting insight into how such mechanisms work in international organizations.

7. Sources and literature

7.1 Sources

In this section the following abbreviations apply:

BT = BT Archives London

PRO = Public Records Office, Kew

Telia = Televerket's archives, held at Landsarkivet, Uppsala

7.1.1 Unprinted

Arendt (1927), Letter from Arendt to GPO, dated Berlin, 21 April, 1927, *BT, Post 33/4703*

AT&T (1934), Memorandum from the AT&T Overseas Services, dated New York, January 3, 1934, *BT, Post 33/4704*

CCI list of urgent international circuits, *Telia, Administrativa byrån/Ekonomi- och Kanslibyrån, F IVb: 42:1*

Comité Consultatif International des Communications Téléphoniques à Grande Distance, Permanent Commission, Second Conference (held in Paris, November 24th-December 1st, 1924), Minutes of proceedings, English version, *PRO, HO 257/51*

Comité Technique Preliminaire 12/3 1923, *Telia, Administrativa byrån/Ekonomi- och Kanslibyrån, F IVb: 42*

Ministry of Foreign Affairs (1938), Instructions to the United Kingdom delegation to the Telegraph and Radiocommunication Conferences of Cairo, 1938, *PRO HO 257/99.*

Post Office (1887) memorandum dated 12 November 1887, regarding whether a Resolution in the House of Commons would be necessary or not, regarding agreements with Belgium and France over the working of the telegraphs to those countries. *BT Archives Post 30/925C*

Post Office (1897), 1891-7 London - Paris telephone service: report on first year of working, *BT Post 30/675 E*

Post Office (1904), 1897-1904 Anglo-French telephone service: extension to Provincial Offices, *BT Post 30/1006B*

Post Office (1905), Correspondence between the Post Office and the Treasury respecting the presentation to Parliament of Post Office Statutory Regulations, 1904-1905, *BT, Post 30/925C*

Post Office (1906) Conventions and Agreements with Foreign Countries, memorandum dated April, 1906, *BT, Post 30/925C*

Post Office (1926), Memo on the leasing of transit circuits. *BT Post 87/24*

Post Office (1931) Brief History of the Overseas Telephone Services, May 1931, *BT, Post 87 / 12*

Post Office (1934a), Letters from the Post Office to the Administrations in Melbourne, Cairo, New Delhi, Pretoria, Canada, and Buenos Aires, dated 21 March, 1934, *BT, Post 33/4704*

Post Office (1934b), Continental Telephone Service 1933-34, *BT Post 30/3811B*

Post Office (1936) "Memorandum detailing certain phases of British Radiotelephone Service Procedure", 17 July 1936, *BT, Post 33/4704*

Ray, F. I. (1954) 50 Years of Telecommunication, manuscript for a speech before the Post Office Telegraph and Telephone Society, London 1954, *BT HIC 002/001/0002*

Reichspostministerium (1933), Letter from the Reichspostminister to the Secretary, General Post Office, 11 July, 1933, *BT, Post 33/4704*

Welch, H.G.G. (1933), Letter from H.G.G. Welch, Post Office, to Dr R.W. King, AT&T Co., 20 November 1933, *BT, Post 33/4704*

7.1.2 Printed

Halsbury's Laws of England, Telegraphs and Telephones, part V, vol. 32, 2nd ed., Butterworth & Co., London, 1939

International Telecommunication Convention, Revision of Cairo, 1938

Post Office Telecommunications Statistics, 1949, Post Office, London,

Report from the Select Committee on the Telephone Service, 1921-2; 1921 (191) vii

Report from the Select Committee on the Telephone Service, 1922 (197) vi

SOU 1992:70, Telelag: Betänkande av Telelagsutredningen

Telegraph Act 1863, An Act to regulate the Exercise of Powers under Special Acts for the Construction and Maintenance of Telegraphs. (26 & 27 Vict. c. 112)

Telegraph Act 1869, An Act to alter and amend "The Telegraph Act 1868." (32 &33 Vict. c. 73)

The Attorney-General v. The Edison Telephone Company of London (Limited) Queen's Bench, C. P., and Ex. Divisions. Dec. 20 1880.
Queen's Bench, C. P., and Ex. Divisions. Dec. 20 1880.

7.2 Literature

Aldcroft, D. (1993), *The European Economy 1914-1990*, Routledge, London

Andersson-Skog, L. & Ottosson, J. (1994), Institutionell teori och den svenska kommunikationspolitikens utformning, *Working Papers in Transport and Communication History 1994:1, Dep. of Economic History*, Umeå University and Uppsala University, Uppsala

Arnold, E. & Guy, K. (1986), *Parallel Convergence: National Strategies in Information Technology*, Frances Pinter, London

Baldwin, F. G. C. (1925), *The history of the telephone in the UK*, Chapman & Hall Ltd, London

Bergdahl, J. (1996), *Den gemensamma transportpolitiken*, Department of Economic History, Uppsala University, Uppsala (Diss.)

Chandler, A. D. (1990), *Scale and Scope. The Dynamics of Industrial Capitalism*, Harvard University Press, Cambridge, Mass.

Chapuis, R. (1976), 'The CCIF and the Development of International Telephony, 1923-1956', in *Telecommunication Journal, 43:III*, pp. 184-197

Clarke, A. C. (1958), *Voice Across the Sea*, Harper & Brothers, New York

Codding, G. A. Jr (1952), *The International Telecommunication Union, An Experiment in International Cooperation*, E. J. Brill, Leyden

Coleman, J. S. (1990), *Foundations of Social Theory*, Harvard University Press, Cambridge, Mass.

Dobbin, F. (1994), *Forging Industrial Policy*, Cambridge University Press, Cambridge

Foreman-Peck (1992), The Development and Diffusion of Telephone Technology in Britain, 1900-1940, *The Newcomen Society's Transactions, Vol. 63*, The Newcomen Society for the study of the history of engineering and technology, London

Foreman-Peck, J., (1995), *A History of the World Economy*, Harvester Wheatsheaf, Hemel Hempstead

Foreman-Peck, J. & Millward, R. (1994), *Public and Private Ownership of British Industry*, Clarendon Press, Oxford

Gill, F. (1924), 'European International Telephony', in *The Electrician*, April 25, 1924

Goldmann, K., Pedersen, M. N., and Østerud, Ø. (eds.) (1997), *Statsvetenskapligt lexikon*, Universitetsforlaget, Stockholm

Grimm (1972), *The International Regulation of Telecommunication 1865-1965*, University of Tennessee, (diss.)

Hansen, A. H. (1932), *Economic Stabilization in an Unbalanced World*, Harcourt, Brace and Company, New York

Headrick, D. R. (1981), *The Tools of Empire*, Oxford University Press, Oxford

Heimbürger, H. (1953), *Svenska Telegrafverket IV*, Televerket, Stockholm

Heimbürger, H. (1968), *Nordiskt samarbete på telekommunikationsområdet under 50 år, 1917-1967*, Telestyrelsen, Stockholm

Heimbürger, H. (1974), *Svenska Telegrafverket V:1*, Televerket, Stockholm

Helgesson, C.-F. (1994), *Coordination and Change in Telecommunications*, The Economic Reseearch Institute, Stockholm School of Economics, Stockholm

Helgesson, C.-F. (1995), 'Technological Momentum and the 'Natural' Monopoly', Paper presented at the SHOT 1995 annual meeting

Helgesson, C.-F., Hultén, S., and Puffert, D. (1995), 'Standards as Institutions. Problems with Creating All-European Standards for Terminal Equipment', in Groenewegen, J., Pitelis, C. And Sjöstrand, S.-E., *On Economic Institutions*, Edward Elgar, Aldershot

Hodgson, G. M. (1988), *Economics and Institutions*, Polity Press, Cambridge

Holcombe, A. N. (1911), *Public Ownership of Telephones on the Continent of Europe*, Constable & Co. Ltd., London

Hovi, J. (1992), *Spillmodeller og internasjonalt samarbeid: oppgaver, mekanismer og institusjoner*, Institutt for statsvitenskap, Oslo, (diss.)

Howard, M. (1972), *The Continental Commitment*, Temple Smith, London, 1972

Hughes, T. (1987), The Evolution of Large Technical Systems, in Bijker, W., Hughes, T. & Pinch, T. (eds.), *The Social Construction of Technological Systems*, MIT Press, Cambridge, Mass.

Israelsen, H. (ed.) (1992), *P&Ts historie, band 3: 1850-1927*, Generaldirektoratet for Post og Telegrafvæsenet, Copenhagen

Jacobaeus, C. (1976), *LM Ericsson 100 år, band III: Teletekniskt skapande 1876-1976*, Stockholm

Kahler, M. (1997), *Liberalization and Foreign Policy*, Columbia University Press, New York

Kaijser, A. (1994), *I fädrens spår...*, Carlssons, Stockholm

Kaijser, A. (1995), 'Från uppfinning till globalt system', in Karlsson, M. and Sturesson, L. (eds.), *Världens största maskin*, Carlssons, Stockholm

114

Knight, F., (1924), The Limitations of Scientific Method in Economics, in Tugwell, R. G. (ed.), *The Trend of Economics,* Alfred Knopf, New York

Larsen, E. (1977), *Telecommunications: A History*, Frederick Muller Ltd., London

Laver, M. (1997), *Playing Politics,* Oxford University Press, Oxford

Lee, K. (1996), *Global Telecommunications Regulation: A Political Economy Perspective*, Pinter, London

Lieber, R. J. (1991), *No Common Power: Understanding International Relations*, Harper Collins, New York

Meyer, H. R. (1907), *Public Ownership and the Telephone in Great Britain*, The Macmillan Company, New York

Perry, C. R. (1977), 'The British Experience 1876-1912: The Impact of the Telephone During the Years of Delay', in de Sola Pool, I. (ed.), *The Social Impact of the Telephone*, MIT Press, Cambridge, Mass.

Pitt, D. C. (1980), *The Telecommunications Function in the British Post Office. A Case Study of Bureaucratic Adaption*, Saxon House, Teakfield Ltd., Westmead, Farnborough

Putnam, R. D. (1988), 'Diplomacy and domestic politics: the logic of two-level games', in *International Organization* 42:3 pp. 427-460

Renaud, J.-L. (1990), 'The Role of the International Telecommunication Union: Conflict, Resolution and the Industrialized Countries', in Dyson, K. And Humphreys, P. (eds.), *The Political Economy of Communications: International and European Dimensions*, Routledge, London

Robertson, J. H., *The Story of the Telephone*, Pitman and Sons, London 1947

Savage, J. G. (1989), *The Politics of International Telecommunications Regulation*, Westview Press, Boulder, Colorado

Schmidt, S. K. (1997), 'Sterile Debates and Dubious Generalisations: European Integration Theory Tested by Telecommunications and Electricity', in *Journal of Public Policy, 16:3*, pp. 233-271

Schneider, V., Dang-Nguyen, G. and Werle, R. (1994), 'Corporate Actor Networks in European Policy Making: Harmonizing Telecommunications Policy', in *Journal of Common Market Studies*, 1994:4, pp. 473-498

Sellars, H. G. (1933), 'A Brief Chronology for Students of Telegraphs, Telephones and Posts', *in Telegraph and Telephone Journal*, 1927-1933

Webb, H. L. (1911), *The Development of the Telephone in Europe*, Electrical Press Limited, London

Networks of Telephony:

Central Actors in the CCIF, 1923-39.

Networks of Telephony: Central Actors in the CCIF, 1923-39[1]

The telecommunications system of today has been described as "the world's largest machine".[2] The understanding of this is of course that it constitutes a vast system of interconnected units that work together as a unified whole. This has not always been the case. For the telephone system to develop from the first experimental line in A. G. Bell's workshop into a truly global system, at least two parallel and complementary developments had to occur.

First of all, the technology of telephony had to evolve from its primitive beginnings where speech was barely audible over a twenty-metre wire, into a state where distance was less of an issue and where a large number of connections could take place in the same cable simultaneously. Secondly, given that this technological development took place independently in a large number of geographically and functionally separate areas, there had to be some form of co-ordination to make sure that the parts of the system worked sufficiently well together in order to function as a unified whole.

During the inter-War period (and to some extent up until today) such co-ordination has been carried out by an international organization known as CCIF.[3] Within this organization the member states worked to find common standards and solutions to common problems, while they at the same time acted strategically in order to influence the international order with their own preferences.[4]

The aim of this article is to find out which actors were *central* in the network constituting the CCIF and its expert study groups. Such actors can be assumed to have had a structural advantage in putting forward their positions in the decision making of the organization. In other words, the aim of this study is to find out which actors were more important than others.

---

[1] Special thanks to Jan Ottosson for reading and commenting extensively on this article. Thanks also to Magnus Carlsson, to Heather Heywood at the ITU Archives, and to the Swedish Transports and Communications Research Board for generous financial support.

[2] Karlsson & Sturesson (1995)

[3] The abbreviation stands for *Comité Consultatif International des communications téléphoniques à grande distance en Europe*, and has over time been known under various acronyms such as CCI, CCIF, CCITT, and ITU-T. For the sake of clarity I will however use CCIF consistently.

[4] Jeding (1998)

This article penetrates the strategic action of member states by looking at the *structure* of the co-operation. The recommendations issued by the CCIF, which were to have an enormous impact on the telephone systems of the member states, were studied, suggested and decided on by individuals who represented different organizations and states. By looking at the overall structure of the CCIF and its committees and the positions of individual representatives in this structure, this article analyses which actors were the key players in this game during the first period of the organization's existence, i.e. 1923-39.

This is done through measuring the *centrality* of the individuals, organizations, and states that participated in the CCIF's work. By studying the centrality in two different organizational bodies of CCIF on three different levels of aggregation, this article can give a relatively detailed answer to the question of which states held a dominant position in the CCIF. The centrality of countries is calculated both as an aggregate measure over the whole period, and studied on a year-by-year basis. Thus the results will be able to say something about changes during the studied period as well as which countries were dominant over the period as a whole.

Other issues that are addressed are: Which type of actors dominated the networks? How were the different types of centrality correlated? Was there a relation between countries having a central position in CCIF and having strong organizations representing them? Was there any relation between central individuals and central organizations and states?

The first meeting of what became CCIF took place in 1923. The choice of time period up to 1939 to study is motivated by the fact that the inter-War era was a discrete period in the CCIF's history. During the Second World War much of the European long-distance telephone network was destroyed, and when CCIF took up its work again after the War it was re-organized in a number of ways.

Standard setting in voluntary organizations has traditionally received relatively little attention in institutional economics. Much of the work done on standardisation has been theoretical and based on micro-economics.[5] Most empirical studies of voluntary standards organizations have primarily dealt with modern data.[6] In their studies of the CCIF[7], Schmidt & Werle (1998) find that liberalisation of the telecommunications sector has dissolved the once stable hierarchical national order in telecommunications. This new order has made it more difficult to define unambiguous national positions, and therefore national political influence over the standardisation process has

---

[5] For instance Farrell & Saloner (1988); David (1985); David & Greenstein (1990)

[6] Weiss & Sirbu (1990); Schmidt & Werle (1998)

[7] Or rather its successor: ITU-T.

become limited to cases where complete technical systems with great technical and economic significance are at stake.[8]

Those that have studied technological co-ordination from a historical perspective have generally tended to regard international standardisation as an inherently technological and harmonious process and ignored the political and conflictual side of it.[9] This study could therefore say something new about how institutions were built for the European telephone network in the inter-War era, by recognising that states had diverging interests and investigating which states had an advantage in pursuing their strategies.

This article is organized as follows: this introduction is followed by a section that defines a few concepts that are important for the present study, and discusses the basic assumptions behind the choice of method. The following section gives an outline of how the CCIF was organized and how the decision making processes worked within the organization. In the next section the networks formed by the CCIF's Plenary Assembly and its Committees of Rapporteurs are analysed and discussed with regard to the centrality of their actors. In the last section a concluding discussion is found.

Method

Centrality in networks

Since the 1930's social scientists have used network theory, based on the assumption that the most prominent or important actors usually are located in strategic locations within the network.[10] Centrality is an indicator of to which extent a node is connected to other nodes in the network. *Degree* is a relatively simple and robust measure of centrality which shows the number of nodes directly linked to the node i.[11]

In this context the centrality of countries, organizations, and individuals in the CCIF's Plenary Assemblies and Committees of Rapporteurs[12] is measured as degree. Worth noting here is that the term 'actor' in this context not necessarily refers to an individual. For the purposes of this article the term 'actor' simply denotes a discrete social unit, and can refer to an individual, an organization, or even a state.[13] However, as Jeding (1998) shows, it is

---

[8] Schmidt & Werle (1998), pp. 267-271

[9] For instance Chapuis (1976); Codding (1952; 1984); Valensi (1965); Grimm (1972)

[10] Wasserman & Faust (1994), p. 169; Knoke & Kuklinski (1982), p. 52

[11] Ottosson (1993), p. 26

[12] These bodies and their functions in the CCIF are described below.

[13] This corresponds to the use of the term in Wasserman & Faust (1994), pp. 17-18. For a discussion on some of the problems with regarding collective entities such as organizations or

meaningful to discuss the strategic behaviour of states and organizations as well as individuals in CCIF's work. Here the opportunities for such strategic behaviour are assessed by looking at the structure of the networks constituted by the different members of the organization.

A country or an organization which had a large number of representatives in these bodies is counted as being more central in these networks than one with a smaller number of representatives. The rationale behind this is that through being represented in these bodies, the state or organization is assumed to have been able to influence their decisions in one way or another. On the individual level a person who was member of a large number of Committees or participated in a large number of Plenary Assemblies is counted as being more central, with the assumption that he or she would have had better opportunities for influencing the CCIF's decisions than someone who participated in fewer Committees or Plenary Assemblies.

Ottosson (1993; 1997) reviews a number of historical network studies using different measures of centrality. The evidence seems unclear as to which centrality measure is more reliable - some studies seem to indicate that more sophisticated indexes are preferable whereas other studies show the simple measure of degree to give a truer picture.[14] Ottosson (1993) compares different measures of centrality for the same data and concludes that they tend to show similar results.[15] In Ottosson (1997), degree is used as a measure of centrality for two reasons. First it shows the same general tendency as other, more sophisticated measures. Secondly it is easier to compare with other studies.[16]

The sort of information available when studying patterns in members' participation in the CCIF gives rise to what in network theory is known as an affiliation network. This makes it different from other forms of network studies in two respects. First, the network is not single but a two-mode network, consisting of both a set of actors and a set of 'events' or, in this study, plenary assemblies or committees. Second, the network describes collections, or subsets, of actors rather than simply pairwise ties between actors. Thus connections between actors (the first mode) are based on linkages established through the plenary assemblies or committees (the second mode).[17] This gives rise to some special features both when representing and analysing the data.

Consider for instance the artificial example with two committees with three members each:

---

states as having stable and unitary preferences see for instance Hovi (1992), pp. 18-22, but also the discussion in Jeding (1998), pp. 20-21 and Boli & Thomas (1999).

[14] Ottosson (1993), pp. 36-38; Ottosson (1997), pp. 54-55

[15] Ottosson (1993), pp. 72-73

[16] Ottosson (1997), pp. 54-55

[17] Breiger (1974); Wasserman & Faust (1994), p. 291

4

Figure 1 Membership of two committees

Committee 1	Committee 2
Anderson	Anderson
Carter	Bates
Davis	Carter

Perhaps the most straightforward way of representing such a network is a matrix that records the affiliation of each actor with each event.

Figure 2 Affiliation matrix for the example of four actors and two committees

Actor	Committee 1	Committee 2	Total
Anderson	1	1	2
Bates	1	0	1
Carter	1	1	2
Davis	0	1	1
Total	3	3	6

Each row in the matrix records an actor's affiliation with the committees, so that 1 in the cell [i, j] describes that actor i was affiliated with committee j, and 0 describes that he or she was not. Another feature worth noting is that the row marginal total in the matrix equals the number of committees that an actor is affiliated with. Similarly, the column marginal total gives the number of actors affiliated with a committee.[18]

Another way of representing an affiliate network is with a bipartite graph, i.e. a graph in which the nodes can be divided into two subsets and where all the lines connecting nodes are between pairs of nodes belonging to different subsets (Figure 3).[19]

Figure 3 Bipartite graph of the affiliation network for the example of four actors and two committees

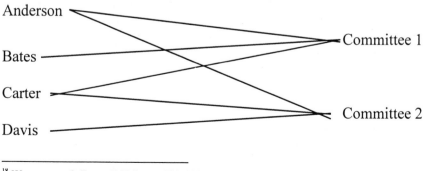

---

[18] Wasserman & Faust (1994), pp. 298-299
[19] Ibid. pp. 299-300

If the network has g actors and h committees, the graph will include $g + h$ nodes. The lines represent the relation 'is affiliated with' (from the perspective of actors) or 'has a member' (from the perspective of committees). From the graph we can see the degree of a node as the number of nodes adjacent to it. In this case the degree of an actor equals the number of committees he or she is affiliated with. Another feature worth noting is that the graph allows us to see indirect connections between actors easier than does the affiliation matrix. Figure 3 shows that Bates and Davis are not connected through co-membership of a committee.[20]

Sources and data

In this study two different affiliate networks are studied. One consists of the members participating in the plenary assemblies (PA's), the other of the members participating in the expert study groups (CR's). For both these networks the centrality of the actors is calculated, measured as degree. These calculations are carried out on three levels: individual members, the organization they represent, and the state they represent.

Since one of the aims in this study is to see which states were dominant during the whole period 1923-1939, the participation of delegates in all the studied Plenary Assemblies and Committees of Rapporteurs are aggregated into two data sets, both covering the whole period.[21]

As showed in Figure 3 above, the degree of an actor in an affiliate network equals the number of 'events' that he or she is affiliated with. Thus on the individual level the degree of an individual actor will equal the number of Plenary Assemblies in which he or she participated in the PA network, and the number of CR's that he or she was a member of in the CR network.

When looking at the organizational and national level, the degree of the actors will instead be aggregate measures. Thus the degree of, say Germany, will equal the sum of all the German delegates that participated in the Plenary Assemblies or Committees of Rapporteurs.

In order to analyse actors' influence over decision making, network studies must focus on the actors' participation in policy events. Knoke (1993) describes such events as critical decision points in a collective decision making sequence that must occur for a collective policy option to be chosen.[22] Also in order to be used, the participation of actors in these events must be known rather than inferred.[23]

---

[20] Wasserman & Faust (1994), pp. 300-301.

[21] For another example of an affiliate network study of a number of actors and a set of events at different points in time, see for instance Breiger (1974).

[22] Knoke (1993), pp. 34-35

[23] Breiger (1974), p. 181; Wasserman & Faust (1994), p. 293

The Plenary Assemblies and the Committees of Rapporteurs meet both these requirements. The Plenary Assembly was the body that made the final decisions on which recommendations the CCIF were to issue, but in order to be put on the agenda of the Plenary Assembly, a draft recommendation had to be suggested by the Committees of Rapporteurs. Therefore both these bodies represented critical points in the CCIF's decision making that had to be passed before a recommendation could be issued.

The result of the analyses can answer a number of questions concerning the position of the members of the CCIF. First they give a measure of which actors were central in the networks respectively. That serves as an indicator of which actors had a structural advantage in putting forward their position when deciding on common European rules.

This reasoning builds on the simple assumption that actors that participated in a Plenary Assembly or a Committee of Rapporteurs had better opportunities for influencing the outcome of the CCIF's decisions than an actor that did not participate. Further it is assumed that an organization or a country that had a large number of representatives in the PA's or CR's had an advantage over organization or country with fewer representatives, *ceteris paribus*, in influencing the CCIF's decision making. In their model of what determines of technological choices in voluntary standards committees Weiss & Sirbu (1990) use participation and representation in the committees as determinants of the outcome.[24]

Another assumption in this study is that individual delegates with a high degree in the networks, i.e. a person who participated in a large number of PA's or Committees, had better opportunities for influencing their decisions than a delegate with a lower degree. This is based on the notion of learning, both of the tacit knowledge of how negotiations and decisions were made, and of the more technical knowledge of the specialised field of international telephony. Another reason for assuming that individuals with high degree had an advantage is that experts with long experience of the CCIF's work could draw on a reputation of expertise and on established contacts with other delegates.[25]

The fact that centrality is measured in the Committees of Rapporteurs as well as the Plenary Assemblies makes it possible to compare the two. To what extent was centrality in one network correlated with centrality in the other? Were the states that invested resources and expertise in the CR's also the ones that sent the most delegates to the Plenary Assemblies? Since centrality is measured in both these bodies, and on three different levels of aggregation (individual, organization, and country) it is possible to give a more fine-grained answer to the question of which states were central in the CCIF.

---

[24] Weiss & Sirbu (1990), p. 128

[25] This is suggested for instance in Fossion (1938) and in the correspondence between delegates. Although well worth a study of its own, this aspect is not specifically studied here.

The data set studied in this article covers the initial meeting of *the Comité Technique Préliminaire* in 1923 and the following plenary assemblies of the CCIF with yearly meetings in 1924-31 and the following conferences in 1934, 1936, and 1938.[26] In the protocols from these meetings are given the names and titles of all delegates to the Plenary Assemblies, as well as their nationality and the organization they represent. In the proceedings are also given the names of the members of the different Committees of Rapporteurs. As far as I have found, these proceedings are the only place where complete lists of members at the PA's and CR's are recorded.

The proceedings from the Plenary Assemblies were sent out to the participating delegations who could comment on them before they went to print and should therefore be regarded as reliable documents. They are kept, with the above mentioned exception of the records from the Plenary Assembly in 1932, in the archives at the ITU headquarters in Geneva. There is also kept a collection of documents by Mr Georges Valensi, Secretary of the CCIF during the period studied here, which gives insights in the early history of the organization.

Participation in the Committees of Rapporteurs was conferred on delegations, who then nominated the individual representative who would take part in the work of the committee. In some cases delegations have failed to nominate a representative, so that in the proceedings a member is simply listed as e.g. 'member of the Italian delegation'. In these cases that delegate is included as an Italian representative at the national level, but excluded at the organizational and individual levels.

CCIF and its decision-making processes

In Europe the telephone systems were built up as national networks, more often than not under the ownership and influence of the national Post, Telegraph, and Telephone Administrations (or PTTs for short).[27] The result of this was that as these national telephone systems developed during the decades around 1900, they came to diverge slightly in technological, economic, administrative, and political aspects.

---

[26] In 1932 a plenary assembly was held in Madrid, co-located with the International Telegraph and Radio Conferences. Since the number and complexity of the questions incited at the previous plenary assembly was too great to allow them to be satisfactory studied in one year, the Madrid meeting was only concerned with questions of tariff structures and exploitation studied by the 6th and 7th CR's. Therefore the protocols for the meeting were not printed, but issued as 'polycopies' to the participants of the conference. (Valensi 1956, p. 51) Copies of those protocols do not seem to be kept in either the ITU archives or the archives of the Swedish or British Telephone Administrations.

[27] For some notes on national differences in the development of national telephone systems, see for instance Jeding, Ottosson & Magnusson (1999), pp.68-9; Andersson-Skog (1997); Noam (1992)

In 1915 a telephone line was established between New York and San Francisco[28], a distance of about 4500 kilometres. This serves to show that technological development at that time, for instance of repeater cables, had made possible telephony over very long distances. In Europe however, there was no way of making a telephone call across the continent. The problem was that the national systems with their different standards, rules, and practices simply were not compatible.[29]

An international telephone call required to be prepared completely from one end to the other before the conversation could start. Thus the telephone systems of the national Administrations concerned needed to be harmonised technically so that the call could be carried between them. The national Administrations also had to agree on minimum standards of maintenance to ensure the reliability of the system. In the days of manual switching precise standards as to how a call should be set up were also needed, in order to minimise time 'lost' in administrating the calls. In addition to this, the application of tariffs and the collection of fees had to be agreed upon in order to establish the international telephone service.[30]

To overcome this problem, the French Ministry of Posts and Telegraphs invited the Administrations of its neighbour countries to a meeting in Paris in 1923.[31] From the foundations laid at that meeting the CCIF was formed with the purpose of working for the creation of an integrated European international telephone system.

Given that the national telephone systems in almost all European countries were strictly regulated and kept under government control, the participating countries were not prepared to give up their sovereignty over the systems in order to form an organizationally integrated European long distance telephone company. Instead a type of cartel agreement was reached where the national systems remained national, but where the CCIF should serve as an international body to provide for unity of direction in means and ends of international telephony in Europe.[32]

The self-appointed task of the CCIF was to standardise equipment and operating methods, to study and select desirable characteristics for long-distance telephone technology, and to plan a programme of work for interconnecting the national systems and developing them into an international network capable of meeting the needs of the day.[33]

The result of the CCIF's studies were issued as recommendations for its member states to follow in their own interest as well as in the interest of the

---

[28] Heimbürger (1974), p. 310
[29] Valensi (1932), p.4
[30] Valensi (1929), pp. 3-4
[31] Comité Technique Préliminaire, 12/3, 1923
[32] Jeding (1998), pp. 40-43
[33] Valensi (1965), pp. 10-11

9

European network as a whole. These recommendations, though institutionally weak, nevertheless proved to have a powerful impact on the ways in which European telephony developed. This further meant that the member states had an incentive to act strategically, in order to suit the common European standards to their own national interests.[34]

Organization of the CCIF

The CCIF's organization built on three different bodies: the Plenary Assembly, the Commissions of Rapporteurs, and the General Secretary. The Plenary Assembly was the executive body. It assembled in a place and at a time decided by the previous PA, and consisted of delegations from those national Administrations or Companies that were members of the CCIF and had stated that they wanted to take part. Only one delegation per country was accepted, so in the cases where several operators took part, these had to be co-ordinated under one head of delegation. The delegations had one vote each in the decisions of the PA.[35]

To make the meetings of the PA more efficient a Permanent Commission was formed, with the task to prepare the next PA, and to study the results of the work of the last meeting. This Commission was "…composed of one member of each Administration of the countries most concerned, either because of the importance of their systems or because of their geographical situation as intermediate countries for through-traffic."[36] In other words, the Permanent Commission was formed of the delegations of those countries whose participation was vital for the interconnection of the European system.

Already at the third Plenary Assembly (in 1926) it turned out that as more states joined the CCIF and wanted to have a say in its studies, the Permanent Commission became impracticably large for its purposes. It was therefore abolished and replaced by specialist Committees of Rapporteurs dealing with the various issues under study and presenting suggestions to the PA who then accepted, modified, or rejected their suggestions. Issues that were not prepared as draft recommendations by a CR could not be put on the agenda of the Plenary Assembly.[37]

In practice the Permanent Commission had already worked in this manner, by assigning different sub-commissions to deal with different aspects of their work. Worth noting is also that in the CR's the national delegates were joined by experts from the equipment industry and international organizations of various kinds. The purpose of this was both to keep operators from

---

[34] See Jeding (1998)
[35] Règlement Intérieur du C.C.I. Téléphonique
[36] Valensi (1929), p. 5
[37] Valensi (1965), p. 11

including impracticable or wasteful provisions in their standards specifications, and to keep the other interested parties informed of the new requirements of operating services.[38]

The PA elected which Administrations or companies were to form the Committees of Rapporteurs, and these delegations designated the individual members of the Committees. In the internal rules of the CCIF it is stated that in order for the CR's to remain efficient, the number of members should be as small as possible, and preferably not larger than six.[39] In practice however, some CR's in some years had over twenty members.

The General Secretary was responsible for co-ordinating the CR's and preparing the following PA. For the sake of impartiality, the G. S. was required not to be in active service of any of the member organizations.[40] For the whole of the period studied here however Georges Valensi of the French Administration filled the office of General Secretary, and the headquarters of the CCIF were located in Paris. The General Secretary also drew up the agenda of the meetings of the Plenary Assembly, based on the reports with draft recommendations from the CR's.[41]

To a large extent the internal rules of the CCIF seem to have been rather informal. The informality of the rules is consistent with the image of the CCIF as a harmonious and conflict free organization devoted to the advancement of communication expressed by several of its early members.[42] Nevertheless there are persistent differences between the positions of different countries on almost all issues, indicating that their work was not simply a matter of joining forces in advancing the technology of long-distance telephony.[43]

One example might serve to illustrate this. At the 1932 Plenary Assembly in Madrid the Italian delegation, supported by Germany, suggested that the organisation of the CCIF should be conformed to that of the International Consultative Committees of Telegraphy and Radio (CCIT and CCIR). The outcome of the International Telegraph Conference was that those two were reorganized to conform to the CCIF. The interesting part is that in an internal memo the General Secretary, Mr Valensi, expresses his certitude that the German delegation supported the Italian suggestion because they had been offended by not being re-elected as hosts of the CCIR.[44] What the General Secretary thus implied was a political motive for action in this allegedly apolitical organization.

---

[38] Valensi (1965), p. 11

[39] Règlement Intérieur du C.C.I. Téléphonique

[40] Règlement Intérieur du C.C.I. Téléphonique

[41] Règlement Intérieur du C.C.I. Téléphonique

[42] See for instance Chapuis (1976); Valensi (1929; 1956; 1965); Fossion (1938)

[43] Jeding (1998)

[44] Valensi (1932), pp. 3-4

Results: The CCIF network

A quick glance at the overall structure of the plenary assemblies shows that the number of individuals participating in them started modestly with 19 delegates at the first meeting in 1923. Over the period the conferences grew into more complex matters, with the number of participants increasing to a peak of 139 in 1936.

Another overview shows the distribution of participants at the plenary assemblies divided by their organizational status (Table 1). In the data set the representatives have been divided into the categories public, private, and organization. The classification refers to which 'mandate' the representative has to the plenary assembly. Thus those labelled 'public' are representatives from a public body, usually the state operator or the ministry responsible for telecommunications, that are members of the official delegation of a country.

The label 'private' stands for representatives of private interests. These were normally parts of the national delegations as well, although their rights to take part in decision-making changed over the period. In countries where there was a public operator, that body was the one with the 'mandate' to send representatives to the CCIF's meetings. However, frequently the public bodies chose to invite representatives from private operators or industry in their countries to take part in the delegations as well.

The label 'expert bodies' finally stands for those whose mandate stems from representing other expert bodies that were invited to take part in the CCIF's work but did not have the right to take part in its decisions. In this category we find bodies such as the International Radio Union, the International Conference of Large Systems of High Tension Electrical Energy etc. These representatives were thus not formally part of the national delegations, but were invited directly by the CCIF.

It is difficult to establish the extent to which, say, a German representative of the International Radio Union was in contact with the German national delegation. Therefore it is even more difficult to ascertain whether or not he would be primarily representing Germany or the International Radio Union. While acknowledging that there is a risk that national interest may have influenced the actions of representatives whose mandate were labelled as 'expert bodies', they have nevertheless been excluded from the national figures in the centrality analyses. Thus the figures in the results section refer only to participants at the plenary assemblies who were part of national delegations. While this is an arbitrary decision, doing the opposite, that is, assuming that 'expert bodies' representatives first and foremost represented their national interest seems even more dubious.

As Table 1 shows, private representation immediately made up just over a fourth of the total number of representatives once private companies were accepted as representatives to the plenary assemblies in 1925. After that, their

Table 1 *Representation at Plenary Assemblies by status*

	1923	1924	1925	1926	1927	1928	1929	1930	1931	1934	1936	1938	Total
Public	19	52	53	65	58	67	83	79	68	80	83	53	760
Private	0	0	21	29	20	4	12	18	27	19	41	11	202
Expert bodies	0	0	8	13	20	22	16	8	20	8	15	2	132
Total	*19*	*52*	*82*	*107*	*98*	*93*	*111*	*105*	*115*	*107*	*139*	*66*	*1094*
% Public	100	100	65	61	59	72	75	75	59	75	60	80	69
% Private	0	0	26	27	20	4	11	17	23	18	29	17	18
% Exp. bodies	0	0	10	12	20	24	14	8	17	7	11	3	12

Source: Protocols from the Plenary Assemblies of the CCIF 1923, 1924, 1925, 1926, 1927, 1928, 1929, 1930, 1931, 1934, 1936, and 1938.

share of the total number of people fluctuated between roughly 10 and 30 per cent, with an overall share of 18%.

The Plenary Assembly in 1928 stands out from the others in having a very low share of private representatives and a somewhat lower total number of delegates. The last meeting on the list, the Plenary Assembly of 1938, also stands out in having only 66 participants and thus breaking the trend of an overall increase in size of the meetings over the period.

Centrality at Plenary Assemblies

One feature of the composition of delegates that might help us identify which actors held powerful positions and thus could be expected to have a relatively large influence on the outcome of the negotiations, is the number of representatives from each country. In Jeding (1998), sending highly qualified engineers to the meetings of the CCIF was identified as one of five main British strategies for influencing the European rules with British preferences.

The voting procedure in the CCIF was one country - one vote. Thus sheer numbers in the delegations did not count for any stronger formal influence in the final decision making. A strong presence at the Plenary Assembly could nevertheless be important for a state that wanted to influence the outcome of the decision making. Logrolling, making deals, and building alliances outside the sessions are all-important tactics in the work of the ITU today[45], and there is no reason to assume that this was different seventy years ago.

---

[45] Shapiro & Varian (1999), ch. 8; Schmidt & Werle (1998), pp. 140-141.

Table 2 *Representation at Plenary Assemblies by country sorted by Total degree*

Country	1923	1924	1925	1926	1927	1928	1929	1930	1931	1934	1936	1938	Total	% Public
Great Britain	4	5	13	16	19	10	11	12	12	10	11	5	128	81
France	4	6	10	11	7	8	9	9	10	12	10	1	97	90
Germany		6	12	14	13	7	11	9	4	5	11	2	94	73
Netherlands		3	4	5	6	7	5	5	5	6	5	2	53	96
Denmark		3	1	2	2	3	6	5	5	3	20	1	51	65
Sweden		4	4	3	3	3	5	4	4	5	4	2	41	100
Belgium	3	3	4	3	2	2	4	5	4	3	4	1	38	95
Italy	2	2	2	2	4	2	2	3	5	5	3	3	35	80
Switzerland	3	3	3	3	3	3	3	3	2	3	3	2	34	100
Hungary		2	4	5	2	1	1	1	0	11	4	2	33	82
Czechoslovakia		3	4	4	3	3	3	3	2	3	3	1	32	94
Argentina								2	8	7	9	5	31	4
Spain	3	1	1	1	2	2	2	4	6	2	3	2	29	83
Austria		1	2	5	2	2	3	3	2	2	2	2	26	92
Mexico							4	6	6	3	5	1	25	9
Romania		1	0	2	1	4	3	2	4	3	3	2	25	52
Norway		3	2	2	2	2	2	2	2	2	4	1	24	96
USSR			3	4	3	4	3	2	2	3	0		24	100
USA			2	0	3	3	6	2	1	3	2		22	0
Poland		1	2	3	2	1	2	0	0	4	5	1	21	95
Japan							3	1	2	2	4	5	17	100
Luxembourg		1	1	2	1	2	2	3	1	0	0	0	13	100
Cuba							2	2	2	2	3	0	11	0
Yugoslavia		1	2	2	0	1	0	0	2	0	0	2	10	100
Finland		1	0	0	0	1	1	0	0	1	3	1	8	75
Portugal			2	0	1	0	0	0	0	2	3		8	100
Egypt											7		7	100
Latvia		2	1	0	0	0	0	2	0	0	0	1	6	100
Chile									2	0	2	1	5	0
Dutch Indies									1	0	0	3	4	100
Estonia			1	0	0	0	0	1	0	0	0	1	3	100
Lithuania			1	1	0	0	1	0	0	0	0	0	3	100
Uruguay											2	1	3	67
China										1	1	0	2	100
Danzig							2	0	0	0	0	0	2	100
Iceland								1	1	0	0	0	2	100
Greece												1	1	100
Mozambique			1	0	0	0	0	0	0	0	0		1	100
Persia							1	0	0	0	0	0	1	100
South Africa										1	0	0	1	100

Source: Protocols from the Plenary Assemblies of the CCIF 1923, 1924, 1925, 1926, 1927, 1928, 1929, 1930, 1931, 1934, 1936, and 1938.

From Table 2 it is striking that for the whole period, Great Britain was with some margin the best represented state at the Plenary Assemblies. Thus measured as *Degree*, Great Britain was the most *central* of the actors on the national level. After Britain come France and Germany as second and third, and then there is a considerable gap to the rest of the states.

Looking at the representation in the Plenary Assemblies on a year-by-year basis shows some variation in the pattern of the most central states. Table 2 shows Great Britain to be approximately in line with France from 1923 and Germany from 1924. After the first years the British dominance becomes more evident. Secondly, the diagram shows the shift in relative strength between Germany and France. From 1924, when they joined the CCIF, Germany remains more central than France at most of the meetings until the early 1930's when their degree suddenly drops. Whether this relatively small representation had to do with national politics or some other factor is not possible to say from the material used in this study, but would be interesting to investigate further.

Another feature of Table 2 is the increasing centrality over the period of non-European countries. Especially delegates from private operators in the Americas give countries like USA, Mexico, Argentina, Chile and Cuba increasingly central positions in the network. To a large extent the delegates from these countries were from companies bought up by the International Telephone and Telegraph Company (ITT). Although the expansion and political ambitions of ITT are well worth a study of its own, the data in this material does not give enough information for a further treatment of that subject.[46]

Figure 4 illustrates the over-all dominance of Great Britain over the whole period, but also the shifts on a year-by-year basis for instance between Germany and France. From the figure the relative decline of the five most central countries is also evident, as more and more countries joined the CCIF.

Another interesting feature is the peak in Danish participation in 1936. Looking again at Table 2 gives other unexpectedly high values for Hungary in 1934, and of Egypt in 1938. That suggests that representation also has something to do with the location of the Plenary Assemblies.

---

[46] See instead Headrick (1991), pp. 201-202; Sobel (1982)

15

Figure 4 Percentage of total representation of the five most central countries in the Plenary Assemblies

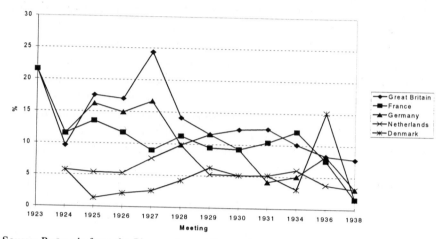

Source: Protocols from the Plenary Assemblies of the CCIF 1923, 1924, 1925, 1926, 1927, 1928, 1929, 1930, 1931, 1934, 1936, and 1938.

Table 3 gives the locations of the relevant Plenary Assemblies. That table gives an indication that states that hosted the conferences tended to be very well represented at the PA's. In fact, excepting the representation of France at the earliest PA's, all states had a larger representation than their over all average at the conferences they hosted. Thus if the conclusion is that it was an advantage for a state to host a Plenary Assembly, then the fact that half of the PA's studied were held in Paris must account for something on behalf of the French Administration. One likely explanation for that fact is that the General Secretary, Mr Valensi, was French, and that the offices of the CCIF were located in Paris.

Table 3 *Locations of the Plenary Assemblies 1923-1938*

Year	Location
1923	Paris
1924	Paris
1925	Paris
1926	Paris
1927	Como
1928	Paris
1929	Berlin
1930	Brussels
1931	Paris
1934	Budapest
1936	Copenhagen
1938	Cairo

Table 4 lists the 15 most central organizations in the PA network; that is, the organizations which had the most representatives at the Plenary Assemblies. The same pattern as in Table 3 is apparent. The British Post Office is by far the most central organization, followed by the French and German national Administrations. One striking feature is that almost all of the most central organizations are public Administrations of the most central countries. The only private company to appear on the list is the American Telephone and Telegraph Company. One other feature worth noting is that whereas Italy is the 8th most central state, there is no Italian organization among the most central 15. The explanation for this is of course that the Italian delegates came from a number of different bodies, none of which alone was central enough to be among the most central. Another example of the same kind is given by the fact that Denmark had an overall degree of 51, but the Danish Directorate General of Posts and Telegraphs had a degree of 30.

The principal interest in measuring the centrality of organizations is that although we can assume that states had national positions that they tried to pursue, a delegation that was made up of representatives from a number of different organizations can be supposed to have had less of a homogeneous position. Jeding (1998) studies how a British position was formed, and concludes that these objectives were formed in a complex game between a number of domestic and international organizations, each with their own set of preferences. As the relative strength of these organizations changed, the British position was shaped and reshaped into what became the actual position in the international negotiations. In this context it is important to note that through this game on the national level, the British delegation had formed a coherent British national position.[47]

National delegations where this formation had not taken place on the national level may have put forward a less unitary set of preferences and given a split impression in the negotiations. It is therefore reasonable to suppose that a state represented by a central organization would be able to pursue a more coherent position than those where the national delegations were made up of a large number of organizations. It can also be assumed that states where the most central organization represented a large proportion of the total number of national delegates had an advantage in putting forward a unitary position. Figures of the proportion of their country's total number of delegates made up by the 15 most central organizations are also given in Table 4.

---

[47] Jeding (1998), pp. 105-107

Table 4 *The 15 most central organizations at the Plenary Assemblies*

Organization	Country	Degree	% of country's degree
British Post Office	Great Britain	104	81
French Ministry of Posts, Telegraphs and Telephones	France	83	86
German Ministry of Posts	Germany	68	72
Dutch National Administration	Netherlands	48	91
Swedish National Administration	Sweden	41	100
Belgian National Administration	Belgium	36	95
Swiss Directorate General of Telegraphs	Switzerland	34	100
Czechoslovak Ministry of PTT	Czechoslovakia	30	94
Danish Directorate General of Posts and Telegraphs	Denmark	30	59
Austrian federal ministry of commerce and communications	Austria	24	92
Soviet People's Commissariat of Posts, Telegraphs and Telephones	USSR	23	96
American Telephone and Telegraph Company	USA	22	100
Hungarian National Administration	Hungary	21	64
Spanish National Administration	Spain	20	69
Norwegian National Administration	Norway	18	75

Source: Protocols from the Plenary Assemblies of the CCIF 1923, 1924, 1925, 1926, 1927, 1928, 1929, 1930, 1931, 1934, 1936, and 1938.

Turning the attention to the centrality of individual representatives gives a somewhat different picture. From Table 5 it is clear that the centrality of states and that of individual representatives did not necessarily go together. The three members who participated in all twelve of the studied Plenary Assemblies represented Belgium and Switzerland, states that were ranked 7th and 9th respectively in order of centrality. One British as well as a French and a German delegate are found in the list, but also representatives from less central states and organizations.

An individual representative with a high degree in the PA network can be assumed to have had a relatively strong position due to his or her[48] experience of the way in which the decision making worked within the CCIF, some cumulated expertise in the rather specialised field of international telephony, and through the development of social networks among the delegates. The correspondence between delegates as well as other records of the dealings of

---

[48] Of all the delegates to the Plenary Assemblies during the period, only two were women. Mrs Dobrouskina, representing the Soviet People's Commissariat of Posts, Telegraphs and Telephones participated in the meetings in 1930 and 1931, and Mrs Babourina of the same organization participated in 1936.

18

Table 5 *Most central delegates at the Plenary Assemblies*

Surname[49]	Organization	Country	Degree
Fossion	Belgian National Administration	Belgium	12
Muri	Swiss Directorate General of Telegraphs	Switzerland	12
Möckli	Swiss Directorate General of Telegraphs	Switzerland	12
Drouet	French Ministry of Posts, Telegraphs and Telephones	France	10
Holmgren	Swedish National Administration	Sweden	10
Höpfner	German Ministry of Posts	Germany	10
Gredsted	Danish Directorate General of Posts and Telegraphs	Denmark	9
Lange	French Ministry of Posts, Telegraphs and Telephones	France	9
Lignell	Swedish National Administration	Sweden	9
Oestreicher	Austrian federal ministry of commerce and communications	Austria	9
Tomits	Hungarian Ministry of PTT	Hungary	9
Trayfoot	British Post Office	Great Britain	9

Source: Protocols from the Plenary Assemblies of the CCIF 1923, 1924, 1925, 1926, 1927, 1928, 1929, 1930, 1931, 1934, 1936, and 1938.

the CCIF speak of rather closely knit personal relations between them.[50] Being part of such a social network must surely have been an asset in the CCIF's negotiations when it came to the informal logrolling and alliance building that most likely were a part of the organization's decision making.

To some extent the argument about having a homogeneous position with unitary preferences, put forward in relation to the centrality of organizations, can be assumed to have had some influence regarding the individual representatives as well. A delegate who, like for instance Mr Fossion, took part in all the studied Plenary Assemblies is likely to have given a consistency to the Belgian position that would have been more difficult to achieve for a state whose delegates alternated more often.

Centrality in the Committees of Rapporteurs

Apart from being central at the Plenary Assemblies a country or an organization could also gain influence by supplying experts to the Commissions of Rapporteurs. If the opportunity to form alliances and trade influence was

---

[49] All the delegates' surnames have been controlled against their given names and titles so as to avoid two people with the same surname being counted as one.

[50] Fossion (1938) writes for instance: "The author of these lines has been shown the honour of participating, since the beginning, in the work of the new organization. ... He has preserved a living memory of the activity, the high competence, the spirit of solidarity and devotion of its founding members." (p. 339) Mr Fossion then goes on to "single out for a table of honour" the names of some of the long-standing members of the CCIF.

apparent at the PA's then that must have been even more so at the meetings of the CR's. Whereas the task of the Plenary Assembly was to accept, reject, or possibly modify the draft recommendations put forward by the CR's, the latter had the opportunity to have a more profound influence on the CCIF's agenda.

Table 6 *Centrality in Committees of Rapporteurs by country sorted by Total degree*

Country	1923	1924	1925	1926	1927	1928	1929	1930	1931	1934	1936	1938	Total
Great Britain	4	2	8	10	8	10	10	9	9	15	13	4	102
France	3	1	9	5	8	9	9	8	9	9	10	5	85
Germany		2	10	8	7	9	9	7	8	12	10	2	84
Netherlands		1	6	8	3	3	4	3	4	6	6	2	46
Italy	4	1	3	3	3	3	3	3	5	7	7	2	44
Sweden		1	7	4	3	4	4	5	5	4	5	2	44
Belgium	3	1	5	4	1	4	4	3	4	5	5	2	41
Denmark			3	6	1	3	4	4	4	4	4	2	35
Switzerland	3	1	6	3	2	0	3	3	3	4	3	2	33
USA					2	3	6	5	5	5	5	2	33
Czechoslovakia		4	6	0	0	0		3	5	4	4	2	28
Spain	3	0	3	3	1	1	1	4	4	4	4	0	28
Poland		4	4	0	0	0	0	0		5	7	2	22
USSR			5	0	3	3	3	3	0	5		0	22
Romania			6	0	0	0	0	4	4	4		2	20
Norway			6	6	0	0	0	0	1	3	3	0	19
Austria			3	5	1	1	1	1	1	0	3	1	17
Japan										8	7	2	17
Mexico							1	1	3	5	5	2	17
Hungary			4	3	0	0	0	0	0	0	3	2	12
Portugal			5	0	0	0	0	0	0		3	1	9
Cuba						2	1	1	1		2	0	7
Yugoslavia			3	4	0	0	0	0	0	0	0	0	7
Luxembourg			3	3	0	0	0	0	0	0	0	0	6
Uruguay											4	1	5
Argentina							1	1	0	1	0		3
Chile									1	2	0	0	3
China											3	0	3
Estonia			3	0	0	0	0	0	0	0	0	0	3
Latvia			3	0	0	0	0	0	0	0	0	0	3
Lithuania			1	2	0	0	0	0	0	0	0	0	3
Mozambique			2	0	0	0	0	0	0	0	0		2

Source: Protocols from the Plenary Assemblies of the CCIF 1923, 1924, 1925, 1926, 1927, 1928, 1929, 1930, 1931, 1934, 1936, and 1938.

The top of the list in Table 6 looks remarkably similar to that in Table 2. Great Britain is by far the most central state in the network, followed by France and Germany close together in 2nd and 3rd places. By and large the same countries appear at the top of both lists, in roughly the same places. There is also a very strong correlation between the centrality of countries in Table 2 and Table 6 (Spearman's rho = .90, $p < .01$)[51]. This indicates that countries that were 'important' in the formation of common rules for the CCIF tended to be central in both the PA and CR networks.

That tendency is probably further strengthened by the fact that almost all of the members of the Committees of Rapporteurs were also delegates at the Plenary Assemblies. Thus a state with a large delegation to the PA could *ceteris paribus* be assumed to be more likely to be well represented in the CR's as well. But the members of the CR's were not randomly chosen from the overall number of delegates at the Plenary Assembly. Instead the PA elected national delegations that possessed some special expertise in the issues studied by a specific committee, and were willing to take part in the committee's work. These delegations then nominated the individual delegates who were to form the committee.[52]

The explanation for the strong correlation between the two centrality measures is probably more likely to be found in the fact that some countries simply were more willing and able to supply a large number of highly skilled experts to the CCIF's work, and thereby had a greater influence over the organization's work.

Some differences can however be found between the countries' centrality in the two different networks. One is that Denmark and Hungary are less central in the CR network. In the PA network they were 5th and 10th on the list, but in the CR network they are 8th and 20th respectively. Both of these countries had extreme values in Table 2 for the PA's in 1936 and 1934 respectively, as the Plenary Assemblies were held in Copenhagen and Budapest. As is clear from Table 6, their high representation in the PA's in these years did not lead to higher representation in the Committees of Rapporteurs.

One other feature is that the United States and the Soviet Union both were more central in the CR network than at the PA's. USA which was 19th on the list of the PA network was 10th on the CR list. As a contrast Argentina was 12th country on the PA list and 26th in the CR network.

---

[51] The centrality measures used in this study are essentially based on frequency counts that can not be expected to be normally distributed. Therefore Spearman's rho is used for calculating the correlations rather than other measures assuming normal distribution, such as e.g. Pearson's r.

[52] Valensi (1932)

Figure 5 Percentage of total delegates of the 5 most central countries in Committees of Rapporteurs

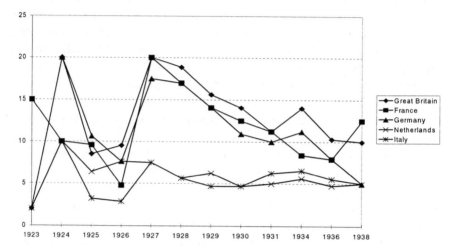

Source: Protocols from the Plenary Assemblies of the CCIF 1923, 1924, 1925, 1926, 1927, 1928, 1929, 1930, 1931, 1934, 1936, and 1938.

Figure 5 shows that the dominance of Great Britain was not established until the 1926 meeting, after which it persisted over the period. The trend over the period in the PA network where Germany was more central than France in the 1920's and fell back during the 1930's is not as strong here. Germany's low degree in the Plenary Assemblies in 1931 and 1934 are not mirrored by their relative strength in the CR network. This implies that although there were fewer German delegates at the Plenary Assembly, their delegates were active in the Committees of Rapporteurs. As in Figure 4 the relative strength of all the 5 most central countries decreases in the 1930's as more countries start participating in the CCIF.

From Table 6 we can further see that the trend recognised in the PA network of increasing centrality for non-European countries is found also in the Committees of Rapporteurs. Thus we see the representation of USA, Mexico, Cuba, Chile, and Argentina increase from the late 1920's. Also the USSR and Japan increase their representation in the CR network over the period.

As in the PA network there are strong similarities between the centrality at national and organizational level. That is, the most central organizations in the CR network tend to be representing the most central countries. Thus in Table 7 we find the British Post Office to be the most central organization, with the French and German public Administrations in 2nd and 3rd place. The difference between the French Ministry of Posts, Telegraphs and Telephones and the German Ministry of Posts is somewhat greater than that between

France and Germany, due to the fact that a relatively larger proportion of the German delegations came from other bodies. The pattern is nevertheless the same as on the national level.

There is a strong similarity between Table 4 and Table 7, that is, the organizations that were the most central in the PA network to a large extent were also the most central in the CR network. This is further indicated by a strong correlation between the centrality of all organizations in the Plenary Assemblies and the Committees of Rapporteurs respectively (Spearman's rho = .63, $p < .01$).

There is also a strong dominance of public telephone Administrations. The only two private companies in Table 7 are the American Telephone and Telegraph Company and the Societatea Anonima Romana de Telefoane.

Quite naturally the most central organizations tend to be those that make up a large proportion of their country's degree. The same discussion as in the section on centrality at PA's above applies here too. That is, a country whose representation was strongly dominated by one organization can be assumed to have had an advantage in putting forward consistent and unitary preferences. That in turn could probably have meant a strategic advantage in the negotiations of the Committees.

Table 7 *The 15 most central organizations in the Committees of Rapporteurs*

Organization	Country	Degree	% of country's degree
British Post Office	Great Britain	100	98
French Ministry of Posts, Telegraphs and Telephones	France	82	96
German Ministry of Posts	Germany	74	88
Dutch National Administration	Netherlands	46	100
Swedish National Administration	Sweden	44	100
Belgian National Administration	Belgium	41	100
Danish Directorate General of Posts and Telegraphs	Denmark	35	100
American Telephone and Telegraph Company	USA	33	100
Swiss Directorate General of Telegraphs	Switzerland	33	100
Czechoslovak Ministry of PTT	Czecho-slovakia	28	100
Soviet People's Commissariat of Posts, Telegraphs and Telephones	USSR	20	91
Norwegian National Administration	Norway	19	100
Spanish National Administration	Spain	19	68
Austrian federal ministry of commerce and communications	Austria	17	100
Societatea Anonima Romana de Telefoane	Romania	14	70

Source: Protocols from the Plenary Assemblies of the CCIF 1923, 1924, 1925, 1926, 1927, 1928, 1929, 1930, 1931, 1934, 1936, and 1938.

Table 8 *Most central delegates in Committees of Rapporteurs*

Surname	Organization	Country	Degree
Fossion	Belgian National Administration	Belgium	17
Gredsted	Danish Directorate General of Posts and Telegraphs	Denmark	17
Höpfner	German Ministry of Posts	Germany	17
Möckli	Swiss Directorate General of Telegraphs	Switzerland	17
Clara Corellano	Compania Telefonica Nacional de Espana/ Spanish National Administration	Spain	16
Breisig	German Ministry of Posts	Germany	15
Collet	French Ministry of Posts, Telegraphs and Telephones	France	14
Holmgren	Swedish National Administration	Sweden	14
Shreeve	American Telephone and Telegraph Company	USA	13
Wiehl	German Ministry of Posts	Germany	13
Bartholomew	British Post Office	Great Britain	11
Lignell	Swedish National Administration	Sweden	11
Nieto	Spanish National Administration	Spain	10
Trayfoot	British Post Office	Great Britain	10

Source: Protocols from the Plenary Assemblies of the CCIF 1923, 1924, 1925, 1926, 1927, 1928, 1929, 1930, 1931, 1934, 1936, and 1938.

On the individual level many of the persons who were the most central in the PA network are recognised also among the most central in the CR network. There is a strong correlation between the centrality of all delegates in the PA and CR networks respectively (Spearman's rho = .52, $p < .01$).

All the most central delegates in Table 8 represented countries that were among the most central in the CR network (see Table 6). It is however interesting that there were no Italian, Dutch or Czech delegates among the most central. Two delegates, Mr Gredsted and Mr Möckli, alone accounted for roughly half of their country's degree. Worth mentioning is also Mr Clara Corellano who shifted from representing the public Spanish National Administration to the private operator Compania Telefonica Nacional de Espana. There are other examples of delegates shifting the organization and even country they represent over time, but none of these were among the most central.

British strategies in the CCIF

The result that Britain was the most central state in the CCIF network is interesting as it gives an indication of Britain's relative strength in pursuing her strategies. Jeding (1998) studies the formation of a British position in the

CCIF, and which strategies she used. That study finds three main factors for explaining which goals that were to become a British position.

First of these is the fact that the British telephone system in the early 1920's was rather primitive as a consequence of under-investment during the first decades of telephony. This had to do with the institutional set up of the British telephone system. The British Post Office jealously guarded its monopoly rights over the operation of a telephone system, but were at the same time unwilling to let the system expand too far since that would threat their revenues from the telegraph monopoly.[53]

The relatively low standard of the British network and a general reluctance on behalf of the British Post Office to invest more than necessary in it, coupled with an insistence that the telephone system should bear its own costs was one force shaping the British position. It led the British delegations to object to standards that would require what they considered too heavy investments ahead of demand. Great Britain was also reluctant to agree to organizational changes in the International Telecommunication Union which, they believed, would lead to increased expenses for the participating states.[54]

A second feature of the British position was the continuous efforts to introduce competition between states in the European telephone network. The tariffs for international calls in Europe was under what was known as 'the CCI system' calculated by adding fixed terminal fees for the countries where a call originated and terminated, and fixed transit fees for the countries through which the call had to pass. Under this system 'standard routes' were also set, indicating which way a call between two countries should normally take. Thus a telephone call for instance from Great Britain to Sweden should follow the standard route through the Netherlands and Germany, unless there was something wrong with that line.

The British position in relation to tariffs was that telephone calls should be priced according to demand rather than cost of production, and also that a system whereby states could compete for transit traffic would be beneficial to the development of the international telephone system in Europe. In this the British position clashed with that of most continental European states, and most notably Germany. The issue of tariffs led to one of the very few examples of open conflicts in the CCIF, where further interconnection between the countries came to a halt in 1926 and 1927 as a consequence of their disagreement.[55]

The opposite views of Britain and Germany over this issue probably have a number of different explanations. One is a general tradition in those countries of different ideas as to what constitutes an efficient organization of eco-

---

[53] Jeding (1998), pp. 60-64
[54] Jeding (1998), pp. 100-101
[55] Jeding (1998), pp. 93-97

nomic or industrial activities. Another explanation is provided by the geographical location of the countries. Great Britain is of course located off the European continent, and before the opening of the Anglo-Irish line, no transit traffic at all went through Britain with the exception of radiotelephony. Germany on the other hand, being a large and central country on the continent, was the most important transit country in Europe. Almost all calls going between Eastern and Western, or Southern and Northern Europe had to pass through Germany. It is therefore not surprising to find Britain more interested in introducing competition between states as a means of lowering transit fees.

Finally, the 'higher' political levels of foreign and defence policy must have played its part in pitting these two countries against each other. Also from this view it must have been an interest of Britain to try to reduce the relative importance of Germany as transit country for such a large proportion of international telephony.

The third and last factor that made the British position different from those of most other European states was a substantially greater interest in extra-European communications. The close relationship with the Dominions as well as with the United States created in Britain a greater demand for communications with the rest of the world, than in most continental European states. This meant two things for the British position. First it gave Britain an interest in developing other technologies than cable technology, which still was of limited use for intercontinental communications. This led Britain to invest in radiotelephony and to become a hub for European radiotelephone communications all over the world.[56]

It also meant that Britain more often than not propagated global rather than European standards. This often implied that the British position in the CCIF was to argue for adoption of the standards used in the more advanced American long distance network, and that the British Post Office suggested the same to the telephone administrations in the Dominions.[57] This view often clashed with that of the continental telephone administrations as well, who often seem to have regarded the CCIF as a European rather than international organization.

Conclusions

One of the primary aims of this study was to find out which states held dominant positions in the CCIF. The answer seems to be that Great Britain with some considerable margin was the most central country, both in the Plenary Assembly and the Committees of Rapporteurs. After Britain there is

---

[56] Jeding (1998), pp. 86-90
[57] Jeding (1998), pp. 90-92

a gap before France and Germany being close in 2nd and 3rd place. Germany, being more central than France in the Plenary Assemblies during the first half of the period falls back in the 1930's to 3rd place. After these three countries there is a group of Western and Central European states together making up a core of central countries in the CCIF, although the centrality of non-European countries increases over the period.

There are some differences between which states were central in the Plenary Assemblies and those that were central in the Committees of Rapporteurs, but on the whole there is a remarkably strong correlation between the two networks. This leads to the conclusion that states that were important in the CCIF tended to participate with a large number of representatives in both these bodies, although exceptions to that rule can be found. Argentina for instance was far more central in the PA than in the CR network, whereas USA held a more central position in the CR's than in the Plenary Assemblies.

Another conclusion to draw from this study is that the most central organizations tended to represent the most central states. Again the British Post Office was by far the most central organization, followed by the French Ministry of Posts, Telegraphs and Telephones, and the German Ministry of Posts. As on the national level there was a strong dominance of organizations from Western and Central Europe, with the American Telephone and Telegraph Company and the Soviet People's Commissariat of Posts, Telegraphs and Telephones to break the pattern.

Also on the organizational level there was a strong correlation between centrality in the two different networks. This indicates that the most central organizations were those that had the resources and expertise to participate extensively in the CCIF's work.

One striking fact about the centrality of organizations is that almost all of them were public telephone administrations, in most cases with national monopolies. Of the 15 most central organizations in the Plenary Assemblies, the American Telephone and Telegraph Company was the only private operator, albeit with a national monopoly on international traffic. In the CR network AT&T and the Romanian private operator Societatea Anonima Romana de Telefoane were the only exceptions.

Looking at the individual level gives a slightly less clear-cut picture of the relation to the centrality of states. There is a strong correlation between the centrality of individuals in the two networks respectively. But the list of the most central individuals does not follow the list of the most central organizations or states as closely. The pattern of Great Britain being by far the most central actor, followed by France and Germany is not repeated. All the most central individuals are however found to represent the core of Western and Central European states, with only one representative from AT&T to break the pattern in the Committees of Rapporteurs.

This study shows that Great Britain devoted plenty of resources in participating in the CCIF, and held a dominant position in the organization. France and Germany were also key actors, although not as central as Britain. This sheds new light on the issue of building institutions for a common European telephone network in the inter-War period. Given that the member states of the CCIF had an interest in influencing the CCIF's decisions with their own preferences, Britain, and to some lesser extent France and Germany, had an advantage in pursuing this. It also confirms the image suggested by earlier studies of CCIF as a rather 'clubby' European organization, dominated by and primarily interested in the European telephone system.[58] The re-organization of CCIF after the Second World War turned CCIF and the International Telecommunication Union in general into a more truly international body. The evidence here however points strongly to the conclusion that before the War, Western and Central European countries had the upper hand in setting the rules for the international telephone system.

Another result is that the most central states also were represented by the most central organizations. Thus the British, French, and German public telephone administrations were the organizations that were in the best position for influencing the CCIF's recommendations. On the whole there was a strong dominance of public monopoly telephone administrations with only few exceptions. This can also say something about the type of recommendations that were likely to come from the CCIF. The public monopoly dominance is a factor that must be taken into account when studying the persistence of central regulation in the field of telephony, and the long-standing view of international telephony as a natural monopoly.

---

[58] See for instance Schmidt & Werle (1998), p. 130

References

Andersson-Skog, L. (1997), The Making of National Telephone Networks in Scandinavia: The State and the Emergence of National Regulatory Patterns, 1880-1920, in L. Magnusson & J. Ottosson (eds.), *Evolutionary Economics and Path Dependence*, Edward Elgar, Cheltenham.

Boli, J. & Thomas G.M (1999), *Constructing World Culture: International Nongovernmental Organizations since 1875*, Stanford University Press, Stanford, 1999

Breiger, R. L. (1974), 'The Duality of Persons and Groups', *Social Forces*, vol. 53:2, pp. 181-190.

Chapuis, R. (1976), 'The CCIF and the Development of International Telephony, 1923-56', *Telecommunication Journal, 43:III*, pp. 184-197

Codding, G. A. Jr. (1952), *The International Telecommunication Union: An Experiment in International Cooperation*, E. J. Brill, Leyden

Codding, G. A. Jr. (1984), 'International Constraints on the Use of Telecommunications: The Role of the International Telecommunication Union', in L. Lewin (ed.), *Telecommunications: An Interdisciplinary Text*, Artech House, Dedham, MA.

Comité Technique Préliminaire, protocols from the meeting held in Paris, *Landsarkivet, Uppsala: Televerket, Administrativa byrån/Ekonomi- och Kanslibyrån, F IV b: 42:1*

David, P. A. (1985), 'Clio and the Econometrics of QWERTY', *American Economic Review*, 75 (2), pp. 332-337

David, P. A. & Greenstein, S. (1990), 'The Economics of Compatibility Standards: An Introduction to Recent Research', *Economics of Innovation and New Technology*, Vol. 1, pp. 3-41

Farrell, J. & Saloner, G. (1988), 'Coordination through Committees and Markets', *RAND Journal of Economics*, Vol. 19 (2), pp. 235-252

Fossion, H. (1938), 'Le Comité consultatif international téléphonique – son origine: son évolution', in *Journal des Telecommunications, 5:XII*, pp. 337-343

Grimm, K. D. (1972), *The International Regulation of Telecommunication, 1865-1965*, University of Tennessee

Headrick, D. R. (1991), *The Invisible Weapon: Telecommunications and International Politics 1851-1945*, Oxford University Press, Oxford.

Heimbürger, H. (1974), *Svenska Telegrafverket V:I*, Televerket, Stockholm

Hovi, J. (1992), *Spillmodeller og internasjonalt samarbeid: oppgaver, mekanismer og institutioner*, Institutt for Statsvitenskap, Oslo

Jeding, C. (1998), 'National Politics and International Agreements: British Strategies in Regulating European Telephony, 1923-39', *Working Papers in Transport and Communications History*, Departments of Economic History, Umeå and Uppsala Universities, Uppsala

Jeding, C., Ottosson, J. & Magnusson, L. (1999), 'Regulatory Change and International Co-operation: The Scandinavian Telecommunication Agreements, 1900-1960', *Scandinavian Economic History Review*, vol. 47, 1999:2, pp. 63-77

Karlsson, M. & Sturesson L. (eds.) (1995), *Världens största maskin*, Carlssons, Stockholm

Knoke, D. (1993), 'Networks of Elite Structure and Decision Making', *Sociological Methods & Research*, vol. 22, No. 1, pp. 23-45.

Knoke, D. & Kuklinski, J. H. (1982), *Network Analysis*, Sage Publications, Newbury Park, Ca.

Noam, E. (1992), *Telecommunications in Europe*, Oxford University Press, Oxford

Ottosson, J. (1993), *Stabilitet och förändring i personliga nätverk: Gemensamma styrelseledamöter i bank och näringsliv 1903-1939*, Uppsala University, Uppsala.

Ottosson, J. (1997), 'Interlocking Directorates in Swedish Big Business in the Early 20th Century', *Acta Sociologica*, Vol. 40 (1), pp. 51-77

Protocols from the Plenary Assemblies of the CCIF 1923, 1924, 1925, 1926, 1927, 1928, 1929, 1930, 1931, 1934, 1936, and 1938.

Règlement intérieur du C.C.I. Téléphonique, *ITU Archives: Depot de M. Valensi*

Schmidt, S. K. & Werle, R. (1998), *Coordinating Technology: Studies in the International Standardisation of Telecommunications*, MIT Press, Cambridge, Mass.

Shapiro, C. & Varian, H. L. (1999), *Information Rules: A Strategic Guide to the Network Economy*, Harvard Business School Press, Boston, Mass.

Sobel, R. (1982), *ITT: The Management of Opportunity*, Sidgwick & Jackson, London

Valensi, G. (1929), *The First Five Years of the International Advisory Committee for Long-Distance Telephone Communications*, Verlag Europäischer Fernsprechdienst, Berlin, *ITU Archives*

Valensi, G. (1932), 'Note sur l'Organisation du Comité Consultatif International des Communications Téléphoniques a grande distance – redigee en vue de la Conference Telegraphique Internationale de Madrid (septembre – 1932.)' *ITU Archives: Dépôt de M. Valensi*

Valensi, G. (1956), *Le Comité' Consultatif International Téléphonique (C.C.I.F.) 1924-1956*, Unprinted manuscript, ITU Library.

Valensi, G. (1965), 'The Development of International Telephony: The Story of the International Telephone Consultative Committee (CCIF) 1924-1956', *Telecommunication Journal 32:I*, pp. 9-17

Wasserman, S. & Faust, C. (1994), *Social Network Analysis: Methods and Applications*, Cambridge University Press, Cambridge

Weiss, M. B. H. & Sirbu M. (1990), 'Technological Choice in Voluntary Standards Committees: An Empirical Analysis', *Economics of Innovation and New Technology*, vol. 1, pp. 111-133.

Regulatory Change and International Co-operation.

The Scandinavian Telecommunication Agreements, 1900-1960.

(Published in Scandinavian Economic History Review, 47:2, 1999.)

Carl Jeding, Jan Ottosson and Lars Magnusson

Regulatory Change and International Co-operation:

The Scandinavian Telecommunication Agreements, 1900–1960

ABSTRACT

The article deals with the issue of international co-operation and co-ordination in the Scandinavian communication industries. While many studies have been made of the international organizations regulating these industries, little has been said about the various voluntary, often informal, bilateral and multilateral agreements that exist outside such organizations. Yet those agreements were influential in shaping the international Scandinavian communications system as well as the national systems. In this article we study the evolution of a Scandinavian forum for co-operation in telecommunications and other network industries during the first half of the twentieth century. One important finding is the path-dependent nature of the co-operation. Once the national authorities had been brought together and their co-operation became fruitful, the field for such action expanded into other issues. Another finding is that as the Scandinavian authorities became more closely connected, they could use this to their advantage in other international organizations.

Introduction

This article examines the evolution of a network of contacts and agreements between the Scandinavian authorities for telecommunications during the first decades of the twentieth century. In doing so, we use the idea of corporate actors to penetrate the nature of international telecommunication co-operation between the Scandinavian countries. Admittedly, it was not the case that any pan-Scandinavian body went against the will of its founders in the respective Scandinavian countries. We wish to stress rather how international co-operation of this kind is 'sticky', path-dependent, which a certain momentum of its own. Once the channels for Scandinavian co-operation were established, their use increased and expanded into new fields of co-operation. Many of these fields were possibly better suited to other forms of regulatory arrangements, but since the Scandinavian authorities already had tried a certain set of arrangements successfully, they were willing to use them also in other fields. In this way, we argue that the choice of organization cannot be understood through static analyses of transaction costs. If we isolate the choice of organization from the particular political and historical context, it simply does not make sense.

Thus we will further develop these questions and discuss the international agreements regarding the network industries, a sector of the economy where politics have influenced economic behaviour in several ways: through regulations, subsidies, direct ownership etc., from the nineteenth century onwards. Our main focus will be on the telecommunications industry, which traditionally has been amongst the most thoroughly regulated areas of national economies.

Given the extensive need for the co-ordination of network industries in general and the telecommunication industry in particular, most European countries have enacted different kinds of regulations. These rules and regulations have been framed in the light of political and economic goals, as well as the needs or wishes of concerned actors. Thus the formal rules regulating the interplay on the communications markets have strongly affected the way in which these markets have worked.[1] The role of a nervous system of information in and between states gives it a strategic importance which by far exceeds its direct economic contribution to states.[2]

For instance, the setting of standards has in this respect mainly been seen as a co-ordination problem; as a process either market driven or initiated by the state through agencies.[3] A recent contribution has shown that the concept of regulatory regimes can be useful in the discussion of co-ordination processes. Hultén and Helgesson view these regulatory regimes as institutions based on the case of the transformation process of all-European standards in the telecommunications sector. They also view the early regimes as 'inward looking'; that is connecting local networks to a national system, so trying to reduce sunk costs.[4]

However, this description lacks an important dimension, namely the making and the role of international agreements in the early, formative period of the telecommunications sector. This dimension raises important questions as to the making and evolution of common rules in an international setting, as well as the relation between formal and informal institutions.

We will argue that the international aspects of co-ordinating the telecommunications systems also inevitably affected national communications policies. Attention will be drawn to the sometimes-neglected form of international co-operation that takes place outside established international organisations. We shall see how the Scandinavian network industries and state agencies acted – and reacted – in the formation of such international agreements by focusing on three aspects: technological links, economic issues (that is governance and the regulatory bodies) and the rise of a common Scandinavian 'position' in the international organisations.

In using the example of Scandinavian co-operation in the field of telecommunications we discuss how forces exogenous to the regulatory process, such as technological developments and the more general questions of international politics,

---

1 Andersson-Skog, L. and Ottosson, J., *Institutionell teori och den svenska kommunikationspolitikens utformning*, Uppsala: Working Papers in Transport and Communication History 1994:1. Department of Economic History, Umeå University and Uppsala University 1994.

2 Foreman-Peck, J. and Müller, J., *The Changing European Telecommunications Systems*, Ed. J. Foreman-Peck, & J. Müller, European Telecommunication Organisations. Baden-Baden: Nomos 1988.

3 Hultén, S. and Helgesson, C.-F., Standards as Institutions. Problems with Creating All-European Standards for Terminal Equipment, in *On Economic Institutions. Theory and Applications*, Eds. J. Groenewegen, C. Pitelis, and S-E. Sjöstrand. Aldershot and Brookfield: Edward Elgar 1995, 172.

4 Ibid.

brought the Scandinavian telecommunications authorities together. Overlapping interests together with a developing social network between the actors led to a close relationship between them, that furthered their co-operation.

Such a relationship, however, did not take the form of an institutionalised organisation. This itself raises several interesting issues. First, it extends the discussion of formal and informal institutions, since the agreements made were often on an *ad hoc*, voluntary basis. Secondly, it deals not only with the question of the actions of any one state, but also with multilateral agreements. Thirdly, it highlights the specific role of national agencies in the related network industries. And lastly, it deals with the role of the network industries themselves, as well as their suppliers, in the process of international co-operation.

Theories of international co-operation in communications

The process of regulation and deregulation in an economy is an important area where politics and economics meet. However, the policy process of the shaping of agreements and regulatory orders between nation states has been rather neglected in both the new and the old institutional economics. Such matters also seem to be overlooked in the traditional regulatory economics within the industrial organisation tradition.

To both understand and fruitfully analyse the actions of the state in the economic sphere, it is necessary to put the formation of economic and regulatory policy into an institutional and historical context. Simply assuming that state intervention is devised in order to support overall optimal solutions, or to maximise the powers of the state for its own ends, defines away exactly the factors that need to be studied. Institutions, the historical context, and, in a broad sense, culture, are factors that set limits to the scope of choice in policy making. By delineating and understanding these limits, we can reach a better comprehension of the regulatory political processes.

This itself raises two important questions. First, in order to understand the motives of acting on behalf of the state we must acknowledge the possible existence of 'national styles', paradigms, and specific regulative orders which thrive upon the existence of path-dependency and enforcing further path-dependency.[5] Secondly, we must also discuss how the interaction between states in international agreements can affect these national styles of institutional settings.

There is a body of theories, developed in the area of international politics, which seems to be more suitable than traditional regulatory economics to explore further the reasons behind international co-operation. Three main schools of thought have dominated the discussion of the role and function of international political bodies: the realist or neo-realist school, the neo-liberal school, and, lastly, the policy network approach.[6]

---

5 Rutherford, M., *Institutions in Economics: The Old and the New Institutionalism*. Cambridge: Cambridge UP 1994, 168; Dunlavy, C. A., *Politics and Industrialization: Early Railroads in the United States and Prussia*. Princeton, NJ: Princeton UP 1994; Dobbin, F., *Forging Industrial Policy: The United States, Britain, and France in the Railway Age*. Cambridge: Cambridge UP 1994; *Evolutionary Economics and Path Dependence*, Eds. L. Magnusson & J. Ottosson. Cheltenham: Edward Elgar Publ. 1997.

6 Lee, K., *Global Telecommunications Regulation: A Political Economy Perspective*. London: Pinter 1996.

Within the neo-realist approach, international co-operation in telecommunications is seen as a struggle between states for achieving strategic high technology, where the various power blocks, such as North America, Europe, and Japan, try to arrange the world's telecommunication order to suit their own interests.[7]

In liberal and neo-liberal theories, such as the functionalist approach, institutions are included in the analysis. The basis for these theories is that international co-operation exists because of the overlapping interests of states. Whether an activity is regulated regionally, nationally, or internationally depends on its function. As the telecommunications industries became ever more international, it was natural that their regulation also became more international. The process of international co-operation may be seen as a process in which a number of factors, such as regimes (concrete procedures and rules of decision making) and conventions (informal institutions) shape the outcome.[8]

Several researchers have lately used the concept of corporate actors to discuss the case where the actors, such as states, transfer some of their authority into a common body.[9] This may be done to create a stable basis for collective action, in order to achieve a common good. Once established, however, these bodies start pursuing their own preferences and aims, which may not be always compatible with those of their original founders. They start living a life of their own, as it were.[10]

In the field of telecommunications, Schneider, Dang-Nguyen, and Werle have applied the corporate actor concept to analyse European telecommunications policy as a multi-level game, where the EU institutions are seen as players in their own right, at a supra-national level.[11]

One area of international co-operation overlooked by these bodies of theory is the kind of voluntary, 'informal' agreements that occur outside formal organisations. In such agreements there is no separate body created to act in the interests of the various actors. Yet, we will argue, the act of co-operation tends to reinforce itself.

Forces exogenous to the regulatory process, for example, technological developments, alter the conditions under which the regulations need to operate, which creates an opportunity to change the institutional framework of the network industry.[12] As different actors come together to solve the new problems of co-ordi-

7 Lee, *Global Telecommunications Regulation*; Arnold, E. and Guy, K., *Parallel Convergence: National Strategies in Information Technology*. London: Frances Pinter 1986.

8 Lee, *Global Telecommunications Regulation*; Renaud, J-L., The Role of the International Telecommunication Union: conflict resolution and the industrialized countries, in *The Political Economy of Communications: International and European Dimensions*, Eds. K. Dyson & P. Humphreys. London: Routledge 1990.

9 Coleman, J. S., *Foundations of Social Theory*. Cambridge, Mass.: Harvard UP 1990.

10 Krueger, A. O., The Political Economy of Control: American Sugar, in *Empirical Studies in Institutional Change*, Eds. Lee J Alston, Thráinn Eggertsson and Douglass North. Cambridge: Cambridge UP 1996; Bergdahl, J., *Den gemensamma transportpolitiken*, Uppsala Studies in Economic History 40, Department of Economic History, Uppsala University 1996.

11 Schneider, V., Dang-Nguyen, G. and Werle, R., Corporate Actor Networks in European Policy Making: Harmonizing Telecommunications Policy, *Journal of Common Market Studies*, 1994:4, 473–498.

12 For a discussion of the effects of technological change on organisation in a more modern context, see Foreman-Peck, J. and Muller, J., The Changing European Telecommunications Systems, *European Telecommunication Organisations*, Eds. J. Foreman-Peck & J. Muller. Baden-Baden: Nomos 1988.

nation, such co-operation places emphasis on their overlapping interests. As co-operation proceeds, the development of a social network between the actors further strengthens the process and provides a force for perpetuating it.

This is not to say, however, that institutional change is a spontaneous process that reacts automatically to its environment. Neither does this imply that the shape of the new set of institutions can be predicted from the change in the environment initiating the institutional change. We suggest that the same environmental change can lead to a wide range of different shapes of institutional outcomes, and that which of these is chosen is determined by specific factors in each case.

The paper is further organised as follows: after an overview of some of the recent discussion concerning international agreements, the case of the Scandinavian telecommunications agreements is discussed in that context, and then the case of Scandinavian co-operation in civil aviation. Certain characteristics of those industries are then compared with developments in related network industries. Some conclusions will then be drawn.

The making of the Scandinavian network regulations

Setting out our analysis with the aim of introducing factors such as national styles and path dependency, we first need to define the context in which the co-operation took place. The development of the electrical telegraph opened up new fields of communication and brought new opportunities of communication between countries. Soon after its introduction, both companies as well as national governments started to build the infrastructure needed for the new technology. Already in its earliest phases, the telegraph pointed to the need for political agreements. New patterns of co-operation arose with the signing of bilateral and multilateral agreements concerning the construction of cable links between countries. The patterns of such agreements concerning the telegraph were later influential when international telephone communications were being developed.

In Scandinavia such international co-operation began in 1858. The three director-generals of the telegraph authorities in Sweden, Denmark and Norway discussed the real problem of complex and arbitrary tariffs for telegrams transmitted through several countries. In subsequent years significant negotiations involving all three countries regarding telegram tariffs took place.[13]

Such negotiations resulted in a generally lower inter-Scandinavian telegraphic tariff, compared with the maximum rates agreed upon in the International Telegraphic Union (ITU). The Scandinavian co-operation, however, was of a rather informal nature, and was not always conflict-free. A proposal from Norway to arrange a Scandinavian conference in 1912, for instance, had to be postponed, due to press reports over accusations from Norwegian officials that the Swedish authorities had tried to prevent Norway from planning a cable to the USA. The director-general of the Swedish Telegraph Authority in fact refused to attend the conference.

Despite these differences, the meeting of the three Scandinavian kings in December 1914, and the neutral status of each of the three countries during the First World War helped the agencies of all three countries to look more positively at

13 Heimbürger, H., *Svenska Telegrafverket, historisk framställning: Det elektriska telegrafväsendet 1853–1902*, Andra bandet. Gothenburg: Elanders 1938.

co-operation. With the introduction of the telephone in the 1880s, the technology, economy and organisation of telecommunications changed. These developments were, of course, interrelated, and successively shaped future developments in that field. We shall outline the forces affecting the Scandinavian telephone networks and their organisation, and how the Scandinavian network evolved.

Structural differences between the Scandinavian countries

The similarities between the Scandinavian countries and their forms of government probably go far to explain the close co-operation between them that evolved.[14] There were, however, some few important differences that need to be explored as a background to this discussion. Perhaps the main difference between the Scandinavian countries was that the Swedish telegraph and telephone administration did not hold a legal monopoly, although it operated a *de facto* monopoly after 1918. In Norway and Sweden separate bodies regulated the postal system and the telecommunications system, whilst in Denmark and Finland from 1927 these functions have been regulated by PTT, post and telecommunications administrations.

Another significant difference regards the market structure of Scandinavian telecommunications, most importantly concerning telephony. The early telephone networks were local businesses, where users formed co-operative-like telephone associations creating local networks, often without any contact with other cities. As the technology advanced, the possibilities for long-distance telephony depended more on organisational factors rather than technology. Telegrafverket (the Swedish public telegraph authority) was initially hesitant about the new technology, but after the use of the telephone extended and began to pose a threat to the telegraph, the authority changed its strategy. From its original position of staying outside the market for telephony, the authority began to purchase several of local networks from the end of the 1880s and connect them into an inter-urban, national network. If we exclude Stockholm, Telegrafverket had by 1902 gained control of over 97 per cent of all the telephones in Sweden. After having bought and incorporated Stockholms Allmänna Telefonaktiebolag (Stockholm's Telephone Co. Ltd.) in 1918, Telegrafverket became in practice the only operator who could offer a telephonic connection to a public network.

In Denmark, in contrast, the public telephone authority, Generaldirektoratet for Post- og Telegrafvaesenet, in these early decades operated the telephone network only in a minor part of the country. The local networks in the rest of the country, which comprised some 95 per cent of all subscribers, were operated by three concession-companies. The national authority was responsible for the traffic between the three concessioned areas and all international traffic.

The Norwegian network structure was essentially similar to the Swedish, and experienced similar developments, although more gradual over time. By the early 1920's there were some 400 private telephone companies in Norway, who controlled about 40 per cent of all telephones. It was not until after the Second World

---

14 For further elaboration on this subject, see Andersson-Skog, L., The Making of National Telephone Networks in Scandinavia: The State and the Emergence of National Regulatory Patterns, 1880–1920, in, *Evolutionary Economics and Path Dependence*, Eds. L. Magnusson & J. Ottosson. Cheltenham: Edward Elgar 1997. Rafto, T., *Telegrafverkets historie, 1855–1955*, Bergen: John Griegs boktrykkeri 1955.

War that the public telephone authority, Telegrafstyret, actively started to purchase the local networks, the process being completed by the early 1970s.

In further contrast, when Finland gained its independence from Russia, all of its telephone network was in private hands, with a special company formed by a number of the local operators operating the inter-urban telephone traffic. This latter company was taken over in 1935 by the public post- and telegraph authority, but the local networks still remained in private hands. Even in the mid-1960s only some 20 per cent of the Finnish local telephone networks were state owned.

Origins of the Scandinavian co-operation in telecommunications

Despite the structural differences that existed between the Scandinavian countries, in telecommunications, the national authorities did decide to co-operate, and once the co-operation was initiated it gradually grew closer as well as encompassing new fields of regulation. In this section we shall study the process of co-operation between the Scandinavian telecommunications authorities, and how that co-operation developed over the twentieth century.

The view that the national state has had a fundamental part in the development of network industries and that they often have been used as instruments in a more general political way is neatly illustrated by the origins of the Scandinavian co-operation in the field of telecommunications. The first organised meeting of the Scandinavian telegraph administrations that is recorded was initiated by the foreign ministries of Sweden, Norway and Denmark, and was held in Copenhagen in 1916. The main issue on the agenda was how make the censorship of telegraphy and telephony more effective in order to hinder the transmission of information that could harm the safety of Scandinavian merchant ships in wartime.

By this date, telephone contact between the Scandinavian countries had long been established, the first official telephone link between Sweden and Norway, for instance, being opened in 1891. Even earlier than this local lines had been established over the national borders, though without official permission. In such cases the lines were purely local affairs without any connection to the national networks.

Continuing negotiations about the inter-Scandinavian telephonic links led to the interconnection of the Swedish, Norwegian and Danish networks in 1893. The extend of communication between them remained limited, however, primarily due to the primitive standard of the national networks. By 1900 a line connecting Denmark and Norway, through Sweden, had been established, and by 1902 the yearly volume of inter-Scandinavian calls had reached almost 90000 (see Table 1). The further opening of telephonic links in 1903 between Sweden and Germany through Denmark, and between Norway and Germany through Sweden and Denmark, meant that the national authorities had to reach agreements on the transit fees for these international calls. Such contacts as they had were, however, carried out in an *ad hoc* way, and no multilateral meetings are recorded.

Following the first official meeting of the Scandinavian telegraph administrations in 1916, there were two more in 1917. Although these meetings were originally organised by the ministries of foreign affairs, and their purpose was primarily to deal with political issues, the administrations found it useful to come to-

Table 1. The number of inter-Scandinavian telephone calls, 1893–1902

Year	Sweden–Denmark	Sweden–Norway	Norway–Denmark	Total
1893	313	856	–	1169
1894	6932	4746	–	11678
1895	9571	5542	–	15113
1896	12124	7169	–	19293
1897	12887	11918	–	24805
1898	13520	14472	–	27992
1899	14721	16534	–	31264
1900	25446	23549	1284	50279
1901	32454	32627	4665	69746
1902	42813	40979	6201	89993

Source: Heimburger, H., *Svenska Telegrafverket, I.* Gothenburg: Elanders 1931, 215–221.

gether to discuss other common problems. After the war the national administrations continued to meet on a regular basis, and the meetings began to deal solely with telecommunications issues.[15]

To summarise: the initiative for Scandinavian telecommunications co-operation did not arise from within the industry. It was the result instead of political circumstances during the First World War, leading the Scandinavian governments to take action on the grounds of state security. These initial attempts at co-operation, however soon developed into a more specialised technical co-operation between state agencies. This clearly shows the importance of the political context as well as more cultural factors, by which it was seen as natural to maintain both co-ordination and co-operation in the engineering field of telecommunications. In contrast to spontaneous order-type of explanations, such processes show the importance of political context.

Line construction and transitory traffic

During the First World War the demand for international telephonic links in Scandinavia increased dramatically, so that the number of international calls immediately after the war were 50–100 per cent higher than the pre-war levels. Due to wartime materials shortages, line construction was not able to keep up with demand, resulting in severely overcrowded lines with express calls making up about half of all the calls made.[16]

15 On a few occasions the Scandinavian telecommunications conferences have coincided with postal conferences. In 1929 this met such a fierce resistance from Swedish Telegrafverket that the conference had to be cancelled altogether. Merging the Swedish Post Office and Telegrafverket was at the time an issue on the Swedish national political agenda, and Telegrafverket strongly resisted this with the contention that postal and telecommunications matters had nothing in common. Fear that a common conference would indicate the opposite made Telegrafverket boycott the conference, which was held in 1930 instead.

16 Heimbürger, H. *Nordiskt samarbete på telekommunikationsområdet under 50 år.* Stockholm: Televerket 1968, 41–42.

Such conditions meant that the construction of new lines, and the division of access to the existing international links proved to be the most important agenda issues of the early Scandinavian conferences, after their initial concentration on censorship. The decisions and arrangements pertainong to the telephone line constructions were throughout this stage of co-operation carried through on a bilateral basis. Countries who were not directly concerned in each individual line proposal were, however, kept as closely informed on the planning and proceedings of line constructions as if they were actually directly involved.

The reasons for such close co-ordination were primarily twofold. First, the international telephonic traffic of the Scandinavian countries was mostly internal to themselves. As late as 1966, for example, the percentages of Norwegian, Danish and Swedish international telephony terminating in the other Nordic countries (Scandinavia plus Iceland and Finland) was 81, 80 and 62 per cent respectively.[17] Secondly, the bulk of the international traffic that was not going to any other Scandinavian country normally had to pass through one of them. Prior to the Second World War, it was the case that all international telephonic traffic from Scandinavia had to pass through Germany. Since there was no direct connection to Germany established between either Norway or Denmark, this meant that all Norwegian traffic with the Continent had to pass through either the Swedish–German sea cable or over Denmark via Sweden. To the east a land cable was established linking Sweden to Finland in the early 1920s, but the quality of the network on the Finnish side permitted only traffic with the most northern parts of Finland. When, in 1928, a sea cable was laid in the Baltic sea, the Scandinavian capitals were, for the first time, connected to Helsinki, and through Finland also to the Soviet Union, Estonia, and Latvia.

The issue of a direct cable between Norway and Denmark was discussed at the Scandinavian telephonic conferences after the First World War. No decision was reached, however, nor when it was discussed again in 1931. A lack of investment capital on the part of Norway seems to have been the decisive stambling block, but there was also a reluctance on Sweden's part to lose the transit traffic incomes from all Norwegian international telephony. The issue was solved in a rather direct way in 1940, when the German occupying forces laid a cable to secure direct telephonic contact between Germany and Norway.

At the end of the Second World War, with much of the German telephone network destroyed, the Scandinavian countries found themselves without cable-telephonic connections with the rest of the Western world. In addition to the rebuilding of the cables linking Germany with Scandinavia, a number of new cables were laid, such as that between Denmark and the Netherlands in 1951, between Norway and Great Britain in 1954, and between Sweden and Great Britain in 1960.

Economic co-operation

A very significant step in Scandinavian telephonic co-operation was the establishment of cable contact with North America. Radio-telephony between Great Britain and the United States had been established in 1927, and directly between Scandinavia and the United States immediately after the Second World War. These radio telephone lines though were highly unreliable, which meant that when new

17 Heimbürger, *Nordiskt samarbete*, 40; see also Rafto (1955), 455–457.

amplifying techniques made the operation of direct sea-cables across the Atlantic possible, such developments served as a great and important improvement. The Atlantic cable also brought a new issue into Scandinavian telephonic co-operation, that is common economic liabilities.

The first transatlantic telephone cable was established jointly by AT&T, the Canadian Overseas Telecommunication Corporation and the British Post Office. The Scandinavian countries were jointly offered one channel in this cable against a guaranteed minimum income of transit fees to the British Post Office. As the demand for transatlantic telephony increased with the quality of the services, offered more cables were laid. In these cables the Scandinavian countries again bought rights of use with a shared economic liability. As a way of reducing the costs for transatlantic telephone connections, the Scandinavian countries including Finland placed some of their connections in a common pool, instead of each country having its own direct lines. This arrangement reduced the demand, and thereby costs, for transatlantic telephone lines by some 15–20 per cent.

A similar development occured with regard the access to an earth station for satellite communications. In 1965 the respective Scandinavian administrations, excluding the Finnish, bought a rights of use of the British satellite station at Goonhilly Downs in Cornwall. Later the Finnish authority also participated. As the demand for satellite telephonic connections increased, and the British station was not able to provide enough capacity, the Scandinavian, including the Finnish administrations went one step further in their co-operation, in 1968 deciding to establish a common earth station of their own.[18]

There were other examples of shared Scandinavian telephonic economic liabilities at this time. They included a joint-funded centre for research and development in tele-satellite technology, as well as common offices for the Scandinavian representatives on various international regulatory bodies.

Co-operation in the international fora

When, in the early 1920s, European states began first to organise themselves to overcome the problems of a fragmented market for long-distance telephony, as discussed earlier, the initiative came from France. In 1923, the French ministry for post and tele-communications invited their opposite colleagues in the neighbouring countries, with the notable exception of Germany, to a conference in Paris. At this conference alternative solutions to the co-ordination problem of European telephony were discussed, by the French, Belgian, Italian, Swiss, British and Spanish telephone authorities, in what was called the *Comité Technique Préliminaire pour la téléphonie à grande distance en Europe*. The main alternatives for the organisation of such European co-operation were the three proposals made by Frank Gill in his presidential address to the Institution of Electrical Engineers in 1922. These were:

- A private European company for long distance telephony, which would control all the international traffic in Europe, under concessions from the concerned European states.
- A monopoly company set up by the European states themselves, with their respective telephone authorities as the sole shareholders.

18 Ibid., 122–124.

- A third alternative, which was regarded by Gill as a temporary solution only, in case the member states were not able to agree on the other two, was for the European telephone authorities to form a common body for the study of the problems and needs of European long-distance telephony. This organisation would also co-ordinate the European national networks through the issue of recommendations on technical standards, methods of measurement, operational procedures etc., and was to be able to enforce these recommendations on the participating authorities.[19]

The first alternative was rejected during the meeting: it is supposed to have been an attempt on behalf of the Bell system to gain control over European long distance traffic as well as the American. The second alternative did not receive much support either. None of the telephone authorities present were willing to give up control of their respective national telephone networks to an international long-lines company, even if owned by the European authorities themselves. The third alternative, however, although originally intended as a temporary solution only, was accepted as the organisational way forward.

The creation of such a form of organisation was motivated, above all, by the need for a unified direction of the international network, and by administrative and organisational aspects of the market for telephony. The object of the organisation was said to be to "design large telephone programmes, to achieve unity in methods of construction and operation, and to carry out the standardisation of the materials and the most important constant factors for international long-distance lines." Here the American situation was emphasised, and the conference affirmed that Europe could not afford to fall behind in the development of telephony.[20]

As has been indicated, the Scandinavian countries had already begun such kinds of co-operation between their respective telephone authorities. So when the Comité Préliminaire decided to invite the various telephone administrations to undertake such co-operation more widely across Europe, the Scandinavian authoroties were already organised enough to act together. The following year, in April 1924, the six original member-states of the committee were joined by another fourteen, at a new conference in Paris.[21] Once more the host nation, France, had managed to avoid inviting Germany to the conference. For the Scandinavian countries, however, international co-operation which excluded Germany was of very limited interest, since all their lines to the Continent passed through German territory. The Swedish and Norwegian administrations made this point forcibly in a letter to the French Ministry for Post and Telegraphs, which resulted in an invitation for Germany as well. At this conference the committee was given a more permanent

---

19 Gill, F., European International Telephony, *The Electrician*, 25 April 1924. That Gill's speech had provided the starting point for this meeting is clearly visible in the protocol. Apart from the tribute to Gill and his 1922 speech which was made by the chairman of the conference at its opening, the conference members on several occasions referred to the three alternative solutions suggested by Gill. Comité Technique Préliminaire pour la téléphonie à grande distance en Europe, opening meeting, first plenary meeting 12 March 1923, Administrativa byrån/Ekonomi- och Kanslibyrån, F IV b:42. Telia's archives.

20 Comité Technique Préliminaire pour la téléphonie à grande distance en Europe, closing meeting, plenary session, 20 March 1923, Administrativa byrån/Ekonomi- och Kanslibyrån, F IV b:42. Telia's archives.

21 The fourteen were: Austria, Czechoslovakia, Denmark, Finland, Germany, Hungary, Latvia, Luxembourg, The Netherlands, Norway, Poland, Romania, Sweden, and Yugoslavia.

form, and the new organisation adopted the name *Comité Consultatif Internation-al des communications téléphoniques à grande distance*, or more briefly, the CCI.

Another early example of concerted Scandinavian action within the CCI was when, at its first meeting, the CCI had wanted to adopt a "standard telephone", that is standardised technical specifications for terminal equipment, in its member states. Most of Europe already had a "standard apparatus" due to the fact that International Western Electric was a *de facto* monopolist in many countries. But in Scandinavia they were not, and after Sweden, Norway, and Denmark had all rejected the need for such standardisation, the issue was dropped.[22]

In their relations with the CEPT, the *Conférence Européenne des Administrations des Postes et des Télécommunications*, the Scandinavian authorities initially joined forces in opposing its formation. This body was set up by the PTT administrations in a number of European countries in order to find a forum for specifically European questions within their fields, without having to take up those at the meetings of the CCI as that body gradually became more truly international. Their common thinking behind this was that there already was an international forum for inter-administrational co-operation which was provided by the International Telecommunications Union and its special branch for telecommunications. The Scandinavian telecommunication authorities, being small in comparison to many of their European counterparts, found that the economic and personal resources involved in participation in yet another international organisation were greater than any benefit to be gained from its work. When the organisation was established in 1959, however, the Scandinavian administrations found that the drawbacks of not being part of such an organisation were even higher, whereupon they joined.[23] On many questions considered in the CEPT the Scandinavians have acted as a bloc, taking advantage of their close links and their in many ways similar structures and overlapping interests.

Not only have the Scandinavian telecommunication administrations often submitted common proposals to the relevant international organisations, and in practice acted as a bloc, they have on a number of occasions gone so far as to send one or more common delegates to represent the joint interests of the Scandinavian telecommunications authorities. The main reason for this, of course, has again been the high costs in economic and personal resources of participation in the ever-increasing number of international meetings dealing with various aspects of telecommunications. Such common representations also shows, however, the closeness of the co-operation that has evolved in Scandinavia in these matters.

The Scandinavian telephony co-operation has in fact taken various forms, including as it does intensive consultation and correspondence, joint study visits, as well as common representation in international fora. The core of such co-operation though can be said to be the near-annual Scandinavian and Nordic conferences. Through these different channels of communication a close social network between the officials of the respective telecommunications authorities has evolved. The correspondence carried on between the senior administrators shows that their personal relations were close and friendly. While study of the conferences held

---

22 Propositioner av de svenska och norska delegationerna, Linjebyrån, FI a:442. Telia's archives. In fact the issue of a standardised European telephone was not seriously discussed again until 1995, when the European Union adopted standard specifications for terminal equipment.

23 Heimbürger, *Nordiskt samarbete*, 75–87.

during the first decades of Scandinavian co-operation shows that the sets of people attending such conferences proved very stable over time.[24]

In short, we wish to emphasise that the initiation of the co-operation between the telecommunication administrations was not a spontaneous process. It was rather, as has been shown, a case of the Scandinavian authorities developing and deepening their co-operation after having been brought together for other reasons, including rather ironically, the need for censorship. The technological possibilities for international telephony were present, but different national administrative practices put obstacles in the way of an integrated telephone system. This, in combination with the awareness of telephone administrators that gains were to be made from co-operation, as shown by the American example, created the opportunity for change in the regulatory process, that is Scandinavian co-operation.

We now turn to the question posed earlier, as to why a co-operative 'regime', devised for solving the specific problem of censorship, came to be used for a wide range of other problems, by stating the path-dependent nature of such organisational solutions. Once the form of direct co-operation between the Scandinavian telephone authorities had been tried and found successful, the barriers for making further agreements between them were substantially lowered. The social- personal network stemming from the continuity of intra-Scandinavian meetings further strengthened the links between the authorities.

We may also note a further organisational development in the national Scandinavian telecommunications systems, where they gradually began to develop their administrations in similiar ways. This hint hints at some form of isomorphism, where contact between the authorities led them to the adoption of similar strategies in organising their internal affairs. Closer investigation of such tendensies is, however, go outside the scope of this paper. But for the sake of our argument concerning the development and momentum of the Scandinavian co-operation it is enough to suggest similar organisational developments as a further force behind the deepening and broadening of the co-operation.

In other network industries, Scandinavian co-operation has taken other forms. For instance, in the case of co-operation between the Nordic countries in the field of civil aviation, Finland, Denmark, Norway and Sweden held several conferences during 1918 and 1919 regarding a common Scandinavian air-traffic law. The informal first meetings by private actors were soon taken over by the governments. However, after the holding of an international air-traffic conference in 1919 in Paris, the idea of a common Scandinavian air-traffic law was overtaken by event. Later on, the air-traffic co-operation between Sweden, Norway and Denmark took the form of the semi-private, SAS (Scandinavian Airlines System). Finland's absence from such co-operation apparently stemmed from political reasons during the Second World War.[25]

In the Scandinavian power industry, the integration processes arising from the local, regional and national power networks of power, also led to co-operative alliances between the Nordic countries. One of the first co-operative projects was

24 The authors of this paper have compiled a list of participants at the Scandinavian and Nordic telecommunications conferences between 1917 and 1965.

25 Ottosson, J., The Making of an Scandinavian Airline Company – Private Actors and Public Interests, in *Institutions and Institutional Change in Transports and Communications. during the 19th – 20th Centuries*, Eds. Olle Krantz & Lena Andersson-Skog, Science History Publications. Canton, Mass.: Watson Publishing International 1998.

the construction of power lines linking Eastern Denmark and Sweden in the 1920s, eventually evolving into a wider Nordic co-operation during the 1960s and 1970s, as Kaijser has shown.[26] He points out the joint organisation *Nordel* as an important part of this Scandinavian collaboration of power exchange. Kaijser views this form of collaboration as a diminishing factor on the evolving of 'national styles'. Another area in this respect is of course the Scandinavian co-operation in the field of postal communications.

Conclusions

We have argued that international agreements, especially those outside formal organisations, have played an important – and neglected – part in the process of forming regulatory regimes in the Scandinavian telecommunications industries.

We have especially focused on the phase when the Scandinavian telecommunications systems were formed as publicly-owned monopolies, and state agencies functioned as important regulatory agents. During this period at the beginning of the twentieth century, various national styles were developed. However, at the same time several important international organisations were set up and agreements concluded in order to co-ordinate the activities of these national systems. Not all these agreements, however, were established through international organisations and codified in 'formal rules', since important 'informal' rules, as well as joint ownership, were established between the Scandinavian countries. There were in fact various forms of co-operation between the Scandinavian countries covering a wide range of activities, and varying from informal contacts to strictly-regulated matters as well as joint ventures.

Historically, the relations between the Scandinavian countries have been close, for geographical, political, cultural, linguistic and other reasons. The closeness of these relations has, of course, varied over time, ranging from, at one extreme, actual wars to, at the other, political and monetary unions.[27] During the First World War, the Scandinavian states, who were all neutral, faced a range of similar problems. In December 1914 the Swedish king, Gustav V, took the initiative in inviting his Danish and Norwegian counterparts, Christian and Haakon, for discussion on how to deal with various aspects of neutrality during wartime. From this followed a closer political relationship between the Scandinavian states throughout the war, and what can be seen as a new 'Scandinavianism', leading to a very close foreign political co-operation by the end of the war.

In the field of telecommunications, when contrasted with the more formally-organised European forms of co-operation, the Scandinavian style of co-operation seems the more elusive and harder to pinpoint. The Scandinavian authorities had nevertheless initiated their collaboration and co-operation before the for-

26 Kaijser, A., Controlling the Grid: The Development of High-tension Power Lines in the Nordic Countries, in *Nordic Energy Systems: Historical Perspectives and Current Issues*, Eds. A. Kaijser and M. Hedin. Canton, Mass.: Science History Publications 1995, 509.

27 For illuminating insights into various forms of Scandinavian co-operation during the twentieth century, see for instance, Johansen, H. C., The Danish Economy at the Crossroads Between Scandinavia and Europe, and Sogner, I., The European Idea: The Scandinavian Answer, both in *Scandinavian Journal of History*, Vol. 18, 1993:1; for notes on 'Economic Scandinavism' in the nineteenth century, see Monrad Møller, A., Economic Relations and Economic Cooperation between the Nordic Countries in the Nineteenth Century, in *Scandinavian Journal of History*, Vol 8, 1993:1.

mation of a European international regulatory body, and, moreover, in many instances in the succeeding decades managed to form a coherent bloc within the international regulatory framework.

This could be argued to have resulted from a gradual deepening of the contacts between the Scandinavian telecommunication authorities. It could also be argued that these relationships did not come about spontaneously. Originating in a politically-initiated agreement to regulate the strategic and foreign political aspects of telecommunications, the meetings of the Scandinavian authorities gradually took on a more technical role. We wish to stress as decisive factors in these developments, on the one hand, technological developments, which gave rise to potential gains from co-operation, and, on the other the provision of a form for co-operation, which was modelled on the politically-initiated meetings.

The fact that the Scandinavian co-operation has persisted and deepened, might well be explained by the increasingly gradually more settled, informal institutions of established channels of contact and social networks. Thus, we suggest that these various national systems of technologies, with marked national styles, to paraphrase Thomas P. Hughes, became more integrated through such international agreements.

Another important issue is the differences found between some network-industries, and the growth of various regulatory regimes within the Scandinavian collaboration. In the telecommunications industry, for instance, the co-operation was based on conferences; in the air traffic area, also conferences, held in the late 1910s and early 1920s, which were then replaced in the mid-1940s by a joint, semi-private company; while in the energy field, a common regulatory body for the Scandinavian countries was set up. The emergence of Scandinavian interaction and collaboration in these industries also varied ovar time.

This leads on to a main point that follows form these Scandinavian network regulations: the chosen form of co-operation could not have been predicted from the technological pressures on the institutional framework. Rather the specific historical context, in which the regulatory framework was shaped, determined what form the co-operation would take, as can be seen in the different forms of co-operation occuring in the Scandinavian telecommunications, civil aviation, and electricity industries.

Finally, an important point that has been brought out in this paper is the significant role 'informal' agreements had between the Scandinavian countries, and the part they play in understanding the development of common Scandinavian policies in the international organisations studied.

Liberalisation and Control.

Instruments and Strategies in the Regulatory Reform of Swedish Telecommunications.

(Published in Magnusson, L. & Ottosson, J. (eds.) *Interest Groups and the State,* Edward Elgar, Cheltenham, 2001.)

Liberalisation and Control: Instruments and Strategies in the Regulatory Reform of Swedish Telecommunications

By Carl Jeding

In most of the industrialised world, technological, economic, and ideological changes have led to pressures to liberalise the telecommunications markets.[1] The threat of international competition has led all major industrialised countries to introduce competition in their national telecommunications markets. In most cases this has implied privatising the previously state-owned monopoly operators.

In general, four types of gains from liberalising the telecommunications sector have been expected. First of all, liberalisation is expected to lower telecommunications costs and stimulate the development of new products and services. This, in its turn, would then lead to higher productivity and efficiency throughout the whole economy. Second, it would stimulate growth of the telecommunications sector itself, an argument that becomes increasingly important as the 'digital revolution' seems to keep making telecommunications an ever expanding part of the international and national economies. Third, a liberalised telecommunications market with developed services and low prices would attract international companies that rely on good communications, such as most knowledge-intensive business corporations. Fourth and last, the liberalisation would make domestic telecommunications operators more competitive, so that they can compete successfully on the developing international market.[2]

In this article it is argued that this liberalisation process has not been one of deregulation. In choosing the case of the liberalisation of the Swedish telecommunications market, this article will study how the introduction of

---

1 For an emphasis of technological developments as the driving force behind deregulation, see Koebel (1990); for an emphasis more on the issue of property rights related to deregulation in general, see De Alessi (1980). For an emphasis on economic factors, see for instance Jackson & Price (1994), pp. 1-4

2 Vogel (1996), pp. 29-30; Ioannidis (1998) touches on the motivations behind liberalisation in Sweden.

competition in that market was closely linked to a change in the regulatory system and a shift of emphasis in state regulatory activity, but not to the abandonment or decrease of regulation.

Steven K. Vogel (1996) argues regarding the nature of the wave of regulatory reform that has swept the industrialised world during the last twenty or so years, that what we have witnessed is reregulation, not deregulation. People commonly confuse the introduction of competition in monopoly markets (liberalisation) with the reduction or elimination of Government regulation (deregulation). Those two are not necessarily linked to each other (Vogel 1996, pp. 3-4). This semantic confusion has to do with a deeper, logic one, and that is the assumption that Governments and markets have a zero-sum relationship. An increase in the power of one, the thinking goes, must mean a decrease in the power of the other. In the real world however, there is no such contradiction between more competition and greater Government control (Vogel 1996, p. 3; Samuels 1992).

The Swedish telecommunications market is in fact a striking example of this. After an initial build-up of the telephony in Sweden, where private local initiatives constructed telephone networks with limited geographic range, the State took control of the national telephone system through buying up these local operations and connecting them to the national trunk line system (Skårfors 1997). The resulting State monopoly was however a *de facto* rather than a legal one. There was no law forbidding other operators than the national Administration to start telephone operations should they wish to do so. Thus in order to liberalise the market, i.e. introduce competition, the State had to introduce new regulations to facilitate entry into the market and to hinder abuse of the dominant position by the Administration.

What do we mean by regulation?

In order to discuss fruitfully whether the observed liberalisation has been accompanied by deregulation as well, we must begin by defining what we mean by regulation. Regulation can be defined in a whole range of ways. A narrow definition regards it as a specific set of commands, where a binding set of rules is applied by a body specifically devoted to this purpose. The other end of the scale regards it as all forms of social control or influence, thus including all mechanisms shaping individual or organizational behaviour in the term 'regulation' (Baldwin & Cave 1999, p. 2).

A middle position between those two defines regulation as deliberate state influence. This could cover all state actions designed to influence industrial or social behaviour, and thus range all the way from legislation to actions based on economic incentives (such as taxation or subsidisation), to the supply of information (Baldwin & Cave 1999, p. 2). John Francis restricts this slightly by giving the following definition of regulation: 'regulation occurs when the state constrains private activity in order to promote the public interest.' (Francis 1993, pp. 1-2).

A very specific definition is offered by Clifford Winston, who defines economic deregulation as 'the state's withdrawal of its legal powers to direct the economic conduct (pricing, entry, and exit) of nongovernmental bodies.' (Winston 1993, p. 1263). By inversion that would mean that regulation is defined as the state having those powers. Although the definition is perhaps overly narrow in that only the directing of pricing, entry, and exit are included, the specification that regulation has to do with legal powers is important.

It is an important point of departure in this article that state or government control over an industry or a company is not the same as regulation. State actions are not regulations simply because they are carried out by the state (Majone 1990, pp. 1-2). The state can exert influence or control in a whole range of ways, but in order to analyse this state influence more precisely we need to set up a number of defining criteria which separate regulation from other forms of state control.

First of all, economic regulation concerns economic sectors and markets. This may sound tautological, but that definition excludes a substantial part of all political initiatives and legislation.

Secondly, we regard regulation as actions of a legalistic character. This implies a number of things: on the one hand it means that regulation involves the issuing of a binding set of rules. This separates regulation as something of a commanding nature from other types of incentive-based regimes such as tax breaks, subsidies etc. This further means that regulation requires some sort of control that it is followed, and enforcement. On the other hand the legalistic character implies that regulation is fairly stable over time.

The legalistic nature of regulations further means that they need to be general in regard to actors. Regulation of the telecommunications sector need to apply to the sector as a whole, and its rules should not be actor-specific. Yet we primarily see regulation as something aimed at specific sectors of the economy, which excludes more general forms of legislation, such as for instance competition law in general, from the term regulation.

One further feature of regulation is that it should constitute deliberate actions as opposed to unintended side-effects of other forms of state action. This means for instance that unintended effects of regulation of some other sector of the economy, which also happens to have an influence over the telecommunications market, should not be regarded as telecommunications regulation. Thus we hold that regulation is something which is specifically aimed at expressed goals. This does not imply an over-idealistic assumption that all regulation is introduced out of public interest for some collective

3

greater good.[3] But policy initiatives such as regulation (or deregulation) are bound to be motivated by the actors taking those initiatives. Why regulate, at this time, and in this way? Even if such justifications are not taken at face value, they are nevertheless important for understanding regulations and their objectives.

With the above qualifications and explanations, we could then summarise that for the purposes of this article we shall regard regulation as deliberate state actions, legally based, and aimed specifically at achieving expressed goals in some sector of the economy.

Why regulate?

Regulation theory has come to develop as an important field of research in its own right within a wide range of academic disciplines, from economics, through political studies and sociology, to anthropology. The explanations to how regulations arise, develop, and decline vary greatly among the writers. A common way of grouping these different approaches is however to focus on the motives for and driving forces behind regulation. With that classification scheme, three broad categories fall out; those that focus on public interest, private interest, and institutions (Francis 1993, pp. 1-8; Peltzman 1989; Stigler 1971; Steinmo, Thelen & Longstreth 1992; North 1990; Dixit 1996).

Historically telecommunications have been among the most strictly regulated areas of national economies. The regulation and/or public control over the telecommunications sector has taken various forms in different countries, ranging from strict and elaborate systems of regulation for private operators to full-blown legal and commercial state monopoly.

The motivations for this heavy involvement of the State in the telecommunications sector have varied through different countries and periods of time. A number of common features in most of those motivations can however be identified.[4]

First of all, and perhaps the reason why the State has become involved in the sector in the first place, is that telecommunications have been regarded as a vital national resource for economic and technological development. When the telephone was introduced in Europe during the last decades of the 19th century, it would perhaps have been a logical choice for the telegraph administrations to take an active part in building up the new medium as well. This, however, did not happen since the telephone was not regarded by the

---

3 For an overview of some of the theories on why states regulate markets, see for instance the introduction of this volume, and Baldwin & Cave (1999).

4 Baldwin & Cave (1999) give another taxonomy of rationales for regulating in general. These four broad categories presented here seem better suited to the specific case of telecommunications.

administrations as having any significant interest to them. Not until the use of telephony started to pose a threat to the revenues from their telegraph monopolies did the administrations start taking an active interest (Jeding 1998; Skårfors 1997). The motivations for the present wave of liberalising the telecommunications sector also stress this importance of the sector to the whole economy of states, although the policy implications are different this time.

Infrastructural resources in general have often been considered key resources of this kind, but the relative importance of telecommunications has increased throughout the 20th century. Matters of national security and the importance of having access to a stable system of communications in times of war or other types of crisis also fall into this category (Headrick 1981, chapter 11). Furthermore the sector has held a symbolic value in the sense that a well-developed telecommunications system has been regarded as a sign of being an economically and technologically advanced nation.[5]

Secondly, the telecommunications sector has long been regarded as a natural monopoly. This argument is based on the idea that the economies of scale involved in producing telecommunications services mean that one producer could do it more cheaply than if several operators were to share the market. Another argument in this strand is that competition in telephony would lead to wasteful duplication of network resources. Much has been written on the alleged natural monopoly of telecommunications.[6] One recent study convincingly argued that the natural monopoly of Swedish telecommunications actually was socially constructed by the actors involved in building up the Swedish telephone network during its initial phase (Helgesson 1999). At any rate, if telephony at any stage has been a natural monopoly, most people seem to agree that it is not so any longer.

The third class of arguments relates to the necessity of technological compatibility within the telecommunications systems. These systems are characterised by a strong 'technical linkage'; i.e. they require an extensive amount of co-ordination in order to function as a whole.[7] Such co-ordination is of course more easily achieved if the number of interested and involved parties is minimised.[8]

Fourth and last are the various public service objectives that most governments apply to the telecommunications sector. Access to at least basic telephony services at affordable and unitary rates is in most industrialised

---

5 Helgesson (1995); On 'technological nationalism', see Fridlund (1999), pp. 40-46, 219.

6 for instance by Hultkrantz (1996), and Noam (1992), ch. 2; Helgesson (1999) gives an overview of the history of the concept 'natural monopoly' in relation to telephone services, pp. 331-340

7 Foreman-Peck & Millward (1994), pp. 1-2; Kaijser (1994) discusses the issue of technical linkage.

8 This becomes even more apparent when the co-ordination has to take place on the international arena. See for instance Jeding (1998) and Genschel & Plümper (1997).

countries regarded as a basic right of their citizens. There is of course considerable national variation as to what should be included in these 'basic rights' and what constitutes 'affordable prices'. The point here is however that governments regard the provision of services to sparsely populated areas and to less well-off consumers as something the market forces will not take care of without some amount of regulation or state involvement.

Regulatory change

Against the background of the above, we see two powerful forces at play in the set of rules for the telecommunications market. On the one hand is the strong tradition of rigorous regulation in almost all aspects of the industry. On the other is the strong will and pressure to liberalise the market. As we shall see, these two forces are not necessarily opposed. The crucial question seems to be: how should the regulatory system be reformed in order to achieve the desired liberalisation? The simplest and most straightforward way to deal with that question is to argue that regulations are inherently inefficient, create inefficiencies, and obstruct the market signals. The reduction of regulation will therefore bring about more liberal and efficient markets.[9]

Other writers instead argue that some form of reregulation is needed to liberalise the telecommunications market, and that this initially may lead to an increase in regulatory activity.[10] According to this view, liberalisation of a network industry can be seen as a three-stage process. In the first phase the industry is characterised by monopoly. Since the national monopolists typically are very powerful in this phase, European countries in general have chosen state ownership for them, in order to have more direct control over their actions. As markets are gradually opened up to competition the industry enters phase two. Here the monopoly situation is replaced by a dual system where monopolies in some markets or segments of markets coexist with competition in others. This situation with large asymmetries between competitors, where the large ex-monopolists typically have a great advantage from their vertical integration, calls for an increased amount and intensity of regulations. This 'hump' of regulation is however of a transitory nature, and once the industry has reached a more mature stage of competition the amount of regulation and regulatory activity will shrink away to some minimum of competition law, assisted by self-regulation and co-ordination from the market actors themselves (Bergman, Doyle et al. 1998).

---

9 See for instance Niskanen (1993); Averch & Johnson (1962) do not explicate this in normative terms, but argue that economic regulation of telecommunications is likely to be ineffective.

10 See for instance Bergman, Doyle et al. (1998)

Yet others claim that the specific nature of the telecommunications sector extends the needs for regulation beyond any transitional period of liberalisation (Preissl 1998), or even that the liberalisation movement actually calls for more regulation. The argument departs from the fact that in industries such as public utilities in Western Europe, the issue has usually not been one of deregulation in the sense of removing administrative 'red-tape'. Rather the aim has been to open the markets to competition, or even to create competition in markets where no such has existed for a very long time. In doing this, the first step has been to reshape the old state monopoly into something, which more closely resembles a private operator. But this also implies a risk of creating an actor with powers to behave as a classic profit-maximising monopolist, with all the welfare losses that come with that.

In order to deal with that risk, the new corporatised or privatised operators have been given commercial freedom in a number of respects, while the core activities of these industries have become subject to formal legislative regulation.

> '...(F)rom the lawyer's point of view, it may be suggested that the process of privatization (and the so far rather limited opening-up of markets) has led to a growth rather than a diminution of formal legislative regulation. ... Indeed, it might be suggested that with regard to former nationalized monopolies, competition is only possible through the creation of artificial legal structures rather than through the operation of market forces.' (Usher 1994, p 1).

Steven Vogel suggests that the liberalisation of telecommunications markets has produced a gap between governmental goals and capabilities. While governments have divested themselves of their traditional instruments of influencing the telecommunications sector, their goals have remained the same. This 'capability gap' then has formed a primary impetus for regulatory changes, that is, for introducing new forms of regulation rather than reducing it (Vogel 1996, pp. 25-31).

One problem with this discussion about the effects on the regulatory system of liberalisation is of course the difficulty of finding measures. Clearly measures of the number of people employed in supervising and enforcing the regulations, or the budget devoted to it, are inadequate measures. What is of interest in this context is the extent and intensity of regulations, i.e. the various parts and/or functions of the market are affected, and to what extent. Thus the strategy adopted here will be to look at the changes in the regulatory system in a functional way, in other words to study the desired functions of the regulatory system, and compare whether or not the liberalised system is coupled to more or less of formal regulation.

The Swedish case: two different regulatory systems

The liberalisation process of the Swedish telecommunications market meant radical changes to the principles for state regulation or intervention as the regulatory system for the telecommunications sector changed. The term 'regulatory system' is here meant to include the legal system relating to the telecommunications sector, and the organizational set-up through which the state has tried to influence it.[11] From having had one state-owned operator with a monopoly on equipment and service provision, Sweden's telecommunications market became one of the most liberalised in the world by the mid-1990's (Karlsson 1998, p. 303).

Regulation under state-owned monopoly

When the Swedish telecommunications administration Televerket (or its predecessor Telegrafverket) was established in 1853 it was in the shape of a state-owned public enterprise.[12] This kind of enterprise has traditionally been used in Swedish administration in the communications sector and in some other industries regarded as nationally vital and/or strategic. Organizationally it takes a middle ground in the administrative system. On the one hand it is more independent from the state than a regular civil service authority, both relating to the goals it should achieve and how these are executed. On the other hand its actions are considerably more restricted than those of a state-owned limited liability company (Karlsson 1998, p. 79).

In a number of important respects, Televerket's status as a state owned public enterprise restricted its independence in operational and policy matters, and made it an instrument of the Government for pursuing telecommunications policy. For instance Televerket was not an independent legal subject. It had to follow administrative laws and regulations like civil service authorities, and its decisions could be appealed against to the Government.

Although Televerket had some freedom to decide on its own internal economic matters, all its major and strategically far-reaching decisions had to be taken by the Government or Riksdag. The Riksdag also decided on the annual budgets and investment plans of Televerket, in addition to having influence on the prices of its products and services. This meant that in reality, the Director-General of Televerket was primarily responsible to the Government rather than to the board of Televerket. Moreover, Televerket did not have its own assets, which instead were part of state property

---

11 This definition corresponds to those of Karlsson (1998) and Thue (1995).
12 The term in Swedish is Affärsverk. Karlsson (1998) translates this into state-owned public enterprise, whereas Noam (1992) suggests the term public service corporation.

(Karlsson 1998, pp. 79-80). What gave Televerket some real influence over telecommunications policy was the asymmetric relationship between the administration and the Ministry for Communications, where the latter was very small and almost all the staff and expertise was with Televerket (Ioannidis 1998, chapter 7).

A liberalised regulatory system

During the 1960's and 1970's, telecommunications was almost a non-issue on the Swedish political agenda. Some minor steps towards corporatising parts of Televerket's equipment industry were taken, albeit in a hesitant and non-systematic way. Political concerns were mainly focused on whether or not employment figures in the industry would decrease as a result of changing technology towards more electronic equipment (Karlsson 1998, chapter 3).

The process of liberalising the Swedish telecommunications market gathered pace in the second half of the 1980's, and in 1993 it resulted in a new telecommunications legislation bill. The bill included a new Telecommunications Act, a revised Radiocommunications Act, and a proposal to corporatise Televerket. In the bill, the state's policy objectives for the telecommunications sector were revised. First of all a new policy objective was to create opportunities for efficient competition in the telecommunications markets. Competition should be the instrument with which consumers should get lower prices, higher quality, and a wider choice of services.[13] Also, the state's responsibility for providing universal service to all citizens was expanded. Technological development in the sector had created new opportunities for communication, so that telefax and low-speed modem data communications should now be included in the basic telecommunications services available to all regardless of geographical location.

The corporatisation of Televerket into Telia AB implied a number of structural and administrative changes. The state kept a 100 per cent ownership of Telia, but the new policy objective of introducing competition meant that Telia should act as an independent operator and that regulation should be neutral towards operators. That in turn implied that many of Televerket's responsibilities as the State's policy instrument had to be transferred to an independent regulator, the National Post and Telecom Agency (PTS).

PTS was installed as the Government's 'watchdog' with an overall responsibility for the national telecommunications system. The agency took over from Televerket a number of competences, such as responsibility for

---

13 Telecommunications Act 1993:597, p. 3

9

the national numbering plan, frequency administration, standardisation issues, and representing Sweden in international co-operation.[14]

A new feature of the Telecommunications Act was to introduce a licensing procedure for operators who wanted to offer telephony services to customers. The principle was that all operators who applied for a license should get one. The licenses could however be linked to certain conditions, such as to connect all other subscribers and operators, as well as allowing third party traffic. Interconnection tariffs should be cost-based and 'reasonable', and PTS was given powers to mediate or even make decisions in the case of interconnection tariff conflicts between operators.[15] Other licensing conditions included obligations to convey messages to emergency services, recognise the needs of persons with disabilities, and the needs of the Swedish Total Defence etc.[16]

To summarise this section, the new regulatory system based on the Telecommunications Act of 1993 meant that the State lost its most powerful regulatory tool- control over the market through direct control over the monopoly operator. Instead a new system was introduced, based on new legislation with a new independent regulator, PTS, to oversee the market. At the same time the motives for state control persisted, or even expanded as efficient competition in the telecommunications market was introduced as a new policy objective on top of the already existing such as an efficient telecommunications system, sustainable and accessible during crises and wartime, and accessible to all citizens at affordable prices.

This change of the regulatory structure of Swedish telecommunications implied two things. Firstly, the pledge by the State not to use its ownership of Telia as a means of implementing telecommunications policy, coupled with the stable, or even increased ambition of policy goals formed a capability gap. In other words, if the government divested itself of one important tool to control the telecommunications sector but still wanted to influence it, it would have to find some other, new means of doing so. Hence reregulation. The new regulatory structure with an independent watchdog (PTS) and a new set of legislation was created to fill the gap between the Government's goals and capabilities.

Secondly, this new regulatory structure did not necessarily imply less state control over the telecommunications sector. Theoretically there is no reason why control through an intermediary, such as an independent supervisory agency, should be less effective than direct control over the operators, allowing for agency costs. Indeed the main motivation for initiating the regulatory change was that it would allow a more efficient fulfilment of the policy goals for the sector. But the regulatory reform meant that the Gov-

---

14 SOU 1992:70, ch. 7

15 Telecommunication Act 1993:597, §20, 20 c, 33.

16 Telecommunications Act 1993:597, p. 17 a.

ernment would have to use different tools for its control: formal regulation, bearing in mind the definition in the previous section, rather than direct control.

Regulation under the two different systems

As a next step in the analysis of the two regulatory systems this section deals with how the state tries to achieve its policy objectives under the two different systems. The analysis will start from the policy objectives declared in the Telecommunications Act of 1993, and compare the means through which those objectives were achieved under the state-owned monopoly system and the liberalised system respectively.

The list of regulatory activities discussed in this paper is by no means complete. A number of tasks connected to public goals for the telecommunication have remained stable through the liberalisation process. In a number of cases the only change has been that a particular task has been transferred from Televerket to PTS, where in practice the same employees perform the same job as before. Among such tasks are for instance administrating the national numbering plan, or allocating radio frequencies. Since the purpose of this paper is to compare the differences in carrying out the political objectives for the telecommunications sector under the two different systems, such aspects of state intervention will be left out of the discussion. Worth noting is also that the overview deals primarily with the instruments for involvement or regulation, rather than the effects of it. Whether the objectives are more fully achieved through one set of institutions or another is left out of the discussion.

The policy objectives of the Telecommunications Act of 1993 are expressed in its section 2, and are stated as follows:

> The provisions of the Act aim at ensuring that private individuals, legal entities and public authorities shall have access to efficient telecommunications at the lowest possible cost to the national economy. This implies, inter alia
>
> 1. that anyone shall be able to use, at his/her permanent place of residence or regular business location and at an affordable price, telephony services within a public telecommunications network,
> 2. that everybody shall have access to telecommunications services on equivalent terms, and
> 3. that telecommunications shall be sustainable and accessible during crises and wartime.
>
> The Government or the public authority appointed by the Government may decide that private individuals, legal entities and public authorities shall be ensured access to telecommunications services or network capacity through public procurement.[17]

---

17 Telecommunications Act 1993:597, p. 2

Ensure efficient telecommunications

This objective is of course a complex task and difficult to achieve. It has in part to do with stimulating investments into telecommunications, and to make sure that improvements and technological advances are realised in the national system. It has also to do with ensuring that the national telecommunications resources are compatible and working as a coherent system, given the positive network externalities of telecommunications.

Under the state-owned monopoly system investment into the national telecommunications system was highly politically controlled. Televerket was the only main operator who invested in the Swedish network, and the Riksdag had to approve of its investment plans (Karlsson 1998, p. 80). The unity of the system of course stemmed from the fact that there was only one player operating. The monopoly in terminal equipment was supported by a very rigorous set of regulations. The rule was that all parts of the telephone network, as well as all equipment connected to it, should be owned and maintained by Televerket. It was thus forbidden for subscribers to attach any piece of equipment to the network, or to the equipment provided.[18]

The equipment market was the first segment of the sector to be opened to competition in 1980. Under the liberalised system equipment has to be type approved by PTS before being put on the market, and increasingly the control is reduced to finding unapproved equipment that causes disturbances in the network in some way.[19] This, then, is clearly a case where liberalisation has brought about a decreased amount of regulatory activity while the policy objective has remained the same.

Another issue related to ensuring that the telecommunications network functions satisfactory as a whole is that of interconnection between operators. The Telecommunications Act states that:

> A party supplying telecommunications services ... is liable on request to facilitate interconnection with any other party providing telecommunications services ...
>
> (...) The compensation for the provision of interconnection of telephony services delivered to a fixed termination point shall be fair and reasonable in relation to the performance costs[20]

---

18 This rule went so far as to include such equipment as note pads and stickers. Karlsson (1998), pp. 141-2

19 As from April, 2000, the type approval procedure will be gradually replaced by a system of market control. This system relies even more heavily on self regulation by the industry, and the regulator's role will be to identify products that can cause disturbances in the network after they have been placed on the market.

20 Telecommunications Act 1993:597, s. 20

12

The problem at hand is that Telia owns and operates the largest part of the Swedish telephone network, most importantly the access network, while at the same time operating as a service provider. This vertical integration of the infrastructure with the 'downstream' service sector gives Telia an incentive to charge their competitors a higher price for the access to the network than their own service branch. At the same time it is of course in the interest of the entrant operators to over-emphasise the advantages and under-estimate the costs of the incumbent for providing the access, in the hope that PTS will give them a favourable treatment as new entrants. Finding the 'right' price for these access and interconnection fees is important both for keeping investment in new infrastructure and the utilisation of the existing infrastructure at an optimal level.[21]

In terms of regulation, this means that PTS have to make sure that the operators fulfil the requirement to facilitate interconnection. Most importantly it means that whenever conflict occurs over interconnection tariffs, PTS will have to mediate between the parties, and ultimately decide at which level the tariffs are 'fair and reasonable in relation to the performance costs.' This arguably represents one of the most important tasks of the regulatory authority, and one that has come into existence only with the new regulatory system.

The effect of liberalisation on the extent and intensity of regulation aiming at ensuring efficient telecommunications is thus somewhat mixed. On the one hand the detailed and rigorous regulatory apparatus concerning equipment that can be connected to the network has been removed. Instead the technical coherence of the system is largely upheld by self-regulation, where the regulatory task of PTS mainly consists of issuing type approval of equipment that are put on the European market.

On the other hand a substantial amount of regulation and regulatory activity is dedicated to ensuring that operators facilitate interconnection with other operators in their networks. This is a regulatory task which is of great importance for the allocation of resources to and within the system, and therefore has large consequences for the operators as well as for the telecommunications system as a whole. It is also a regulatory task which, of course, did not exist before liberalisation.

Ensure sustainable and accessible telecommunications during crises and wartime

The level of the Total Defence's needs for telecommunications is ultimately decided by the Riksdag. The State's activity in this respect typically involves constructing protected sites for key elements of the telecommunications system, and carrying out tests of how the system works during manoeuvre

---

21 See for instance Bergman, Doyle et al. (1998), ch. 6

exercises. These activities are funded through the Defence budget. Under the old system Televerket performed all those functions, whereas under the liberalised system PTS follows a procurement procedure where different operators can bid for the contracts under competition. The activities to ensure national needs for telecommunications during crises in peacetime follow the same procedure, but are financed through fees from license holders.

Here too the policy objectives have remained constant. What is new is that PTS procures the actual work from operators, rather than having Televerket do it as under the monopoly system. The carrying out of the procured tasks, such as constructing protected sites for important nodes in the telecommunications system, is a question of operation rather than regulation, regardless of who performs it. This then means that the fact that other operators carry out some of these tasks instead of the state owned operator, does not mean a decrease in regulation. The extent of regulation used to achieve the policy objective of ensuring sustainable and accessible telecommunications is thus rather the same under the two different systems, or has increased slightly with liberalisation as a new procurement procedure has been added.

Ensure universal service

As expressed in the Telecommunications Act, the objective is that anyone should be able to use at his/her permanent place of residence or regular business location and at an affordable price, telephony services within a public telecommunications network; and that everybody shall have access to telecommunications services on equivalent terms.[22]

Here the definition as to what should be included in the universal service has changed over time with technological development. The instrument for ensuring that it is fulfilled has however remained about the same. Under the state-owned monopoly system it was the responsibility of Televerket to fulfil the universal service requirements, and their decisions could be appealed against to the Government. Under the liberalised systemsuch responsibilities can be included in the license conditions of any operator. In fact, and as a pro-competitive measure, Telia is the only operator whose license at this time[23] includes an obligation to provide telephony services to anyone requesting it.

As the market becomes more mature and there is less justification for unilaterally putting the whole burden of supplying services to unprofitable areas, it might be reasonable to devise some system whereby the operators

---

22 Telecommunications Act 1993:597, p. 2
23 January, 2000

share the costs of this. That will then require some further regulatory activity from PTS to allocate those costs between the operators.

Included in the notion of universal service is also that it should be provided at 'an affordable price'. This is a long-standing political objective, which under the monopoly system led to a strong political control over Televerket's tariffs. According to one study, the politicians on Televerket's board were rarely interested in major strategic decisions. They took a strong interest however when tariffs were discussed (Ioannidis 1998, p. 297).

Under the liberalised system price controls are less of a political issue. Subscribers' prices are set independently by the operators. Given the homogeneity of services, at least for standard telephone calls over the fixed network, price competition is by far the most important means of competing for the operators. The government naturally has no power (and no expressed ambition) to use its ownership of Telia to set any specific prices for the subscribers.

The main regulatory task in that respect is to supervise the tariff bases of Telia, and above all their interconnection rates. Here PTS spends some considerable effort in supervising Telia's records to ensure that no cross-subsidies or other abuses of their dominant position is taking place. Also the Competition Authority is involved here, in controlling that the market is characterised by sound competition, and that no cartel agreements are set up.

The relevant section of the Telecommunications Act further states that everybody shall have access to telecommunications services on equivalent terms. That passage is intended to cover the needs of people with disabilities, and implies services such as text telephony, speaking directory enquiries etc. As in the case with defence needs, those services are financed over the state budget, and are procured by PTS under competition.

Ensure competition

Introducing competition in the telecommunications sector is of course the most important change with the new regulatory system, and the very base for it. Section 3 of the Telecommunications act states that:

> (W)hen implementing the Act the endeavour shall be to create scope for and maintain efficient competition within all parts of the telecommunications sector as a means of achieving the objectives specified in Section 2.[24]

A significant part of the Act is devoted to ensuring that an efficient competition is established and upheld in the various segments of the telecommunications sector. For this purpose PTS is given far-reaching powers for su-

---

24 Telecommunications Act 1993:597, p. 3

pervising the license holders, and matters related to promoting competition takes up a large part of the work of the Agency. Naturally much of this work is related to alleged abuse of dominant power by the incumbent Telia. This is one area where the amount of regulation clearly has increased due to the liberalisation of the telecommunications markets and the introduction of a new policy objective.

In summing up this section it might be useful to present the regulatory activities prompted by each policy objective under the two different regulatory systems

Table 1. *Regulatory activities under two regulatory systems*

OBJECTIVE	MONOPOLY SYSTEM	LIBERALISED SYSTEM
Control of terminal equipment	Strictly regulated monopoly	Self regulation. Type approval and market control by PTS
Interconnection	-	PTS supervises and mediates/decides in conflicts
Price control	Strong political influence by the Riksdag/Government	PTS controls that the tariffs are cost-based
Defence needs	Riksdag decides level, Televerket carries out work	Riksdag decides level, PTS procures work
Universal service	Televerket obliged to provide Universal service	Obligation included in licensing conditions
Competition	-	PTS supervision of operators Competition Authority supervises general competition features

Conclusions

The case of the Swedish telecommunications sector seems to justify Vogel's notion of a gap between political goals and capabilities. The liberalisation of the system meant that the State divested itself of its most powerful regulatory instrument - the direct control of the monopoly operator. At the same time all the existing political objectives remained constant and a new objective, the promotion of efficient competition, was added.

This capability gap opened for a change of regulatory systems. First of all it followed that the style of regulation would have to be different. The liberalised regulatory system called for neutral regulations and did not go well together with special relations between the state and Telia. Thus the direct political control of Telia had to be replaced by more general telecommunication legislation, and the regulatory and administrative functions held by Telia were transferred to an independent regulatory agency. In short, the regulatory influence had to shift from being specific and applying to specific issues, into being of a more general nature, applying to all the players on the market.

The instrument through which the state can primarily control the operators under the liberalised system is by conditioning their licenses. Since these licenses, in order to stimulate investment and attract operators, must be transparent and stable over time, that also means that the regulatory influence over the telecommunications market has had to shift from *ex post* to *ex ante*. In an institutional context which is incorporated in a political system that provides safe-guards against legislative and executive opportunism, the regulatory system is more likely to have credibility and attract long-term investments.[25]

Another result, paradoxical to some, is that that the liberalisation does not seem to have given any clear evidence of a reduction of regulations. Instead, we see a whole range of new regulatory measures introduced with the new system, most of these related to the promotion of competition.

If we thus accept Vogel's thesis that the new regulatory system has emerged to fill the gap between political ambition and ability, this leads to the conclusion that deregulation, i.e. the abandonment of regulation, is not likely to occur as long as the policy objectives are constant. Arguably some of the regulation could take different forms. For instance part of the promotion of competition could gradually become less sector specific and be handled by more general competition law. Part of the universal service obligation could also possibly be handled through more general regional policy. The total amount of regulatory activity and market intervention would however remain roughly constant.

---

25 Levy & Spiller (1994)

References

Averch, H. & Johnson, L. L. (1962), 'Behavior of the Firm under Regulatory Constraint', <u>American Economic Review</u>, LII (5), December, pp. 1052-69, reprinted in: Bailey, E. E. & Rothenberg Pack, J. (eds.) (1995), <u>The Political Economy of Privatization and Deregulation</u>, Cheltenham, UK and Lime, US: Edward Elgar.

Baldwin, R. & Cave, M. (eds.), (1999), Understanding Regulation: Theory, Strategy, and Practice, Oxford: Oxford University Press.

Bergman, L., Doyle, C. et al. (1998), <u>Europe's Network Industries: Conflicting Priorities</u>, London: Centre for Economic Policy Research, SNS.

De Alessi, L. (1980), 'The Economics of Property Rights: A Review of the Evidence', <u>Research in Law and Economics</u>, 2, 1-47, Reprinted in: Bailey, E. E. & Rothenberg Pack, J. (eds.) (1995), <u>The Political Economy of Privatization and Deregulation</u>, Cheltenham, UK and Lime, US: Edward Elgar.

Dixit, A.K. (1996), The Making of Economic Policy: A Transaction-Cost Perspective, Cambridge, Massachusetts, London, England: The MIT Press.

Foreman-Peck, J. & Millward, R. (1994), Public and Private Ownership of British Industry 1820-1990, Oxford: Clarendon Press.

Francis, J. (1993), <u>The Politics of Regulation: A Comparative Perspective</u>, Oxford, Blackwell Publishers.

Fridlund, M. (1999), <u>Den gemensamma utvecklingen: Staten, storföretaget och samarbetet kring den svenska elkraftstekniken</u>, Stockholm: Brutus Östlings Bokförlag, Symposion.

Genschel, P. & Plümper, T. (1997), 'Regulatory Competition and International Cooperation', <u>MPIfG Working Paper 97/4</u>, Cologne: Max-Planck-Institut für Gesellschaftsforschung.

Headrick, D. R. (1981), <u>The Tools of Empire: Technology and European Imperialism in the Nineteenth Century</u>, Oxford, Oxford University Press,.

Helgesson, C.-F. (1995), 'Technological Momentum and the "Natural" Monopoly', paper presented at the SHOT 1995 annual meeting, Stockholm: Stockholm School of Economics.

Helgesson, C.-F. (1999), <u>Making a Natural Monopoly: The Configuration of a Techno-Economic Order in Swedish Telecommunications</u>, Stockholm: EFI The Economic Research Institute, Stockholm School of Economics.

Hultkrantz, L. (1996), 'Telepolitikens ekonomiska teori', <u>CTS Working Paper 1996:6</u>, Centre for Research in Transport and Society, Borlänge: Högskolan Dalarna.

Ioannidis, D. (1998), <u>I nationens tjänst? Strategisk handling i politisk miljö</u>, Stockholm: EFI: The Economic Research Institute, Stockholm School of Economics.

Jackson, P. M. & Price, C. M. (1994), <u>Privatisation and Regulation: A Review of the Issues</u>, London: Longman.

Jeding, C. (1998), 'National Politics and International Agreements: British Strategies in Regulating European Telephony, 1923-39', Working Papers in Transport and Communications History, Uppsala: Departments of Economic History, Umeå and Uppsala Universities.

Kaijser, A. (1994), I fädrens spår : den svenska infrastrukturens historiska utveckling och framtida utmaningar, Stockholm: Carlssons.

Karlsson, M. (1998), The Liberalisation of Telecommunications in Sweden, Linköping, Sweden: Department of Technology and Social Change, Linköping University.

Koebel, P. (1990), 'Deregulation of the Telecommunications Sector: A Movement in Line with Recent Technological Advances', in: Majone, G. (ed.) (1990), Deregulation or Regulation? Regulatory Reform in Europe and the United States, London: Pinter Publishers.

Levy, P. T. & Spiller, B. (1994), 'The Institutional Foundations of Regulatory Commitment: A Comparative Analysis of Telecommunications Regulation', Journal of Law, Economics and Organization, vol. 10:2, pp. 201-46.

Majone, G. (1990), 'Introduction', in: Majone, G. (ed.), Deregulation or Regulation? Regulatory Reform in Europe and the United States, London: Pinter Publishers.

Niskanen, W. A. (1993), 'Reduce Federal Regulation', in: D. Boaz & E. H. Crane, Market Liberalism - A Paradigm for the 21st Century, Washington D.C.: Cato Institute.

Noam, E. (1992), Telecommunications in Europe, New York: Oxford University Press.

North, D.C (1990), Institutions, Institutional Change and Economic Performance, Cambridge: Cambridge University Press.

Peltzman, S. (1989), 'The Economic Theory of Regulation after a Decade of Deregulation', Brookings Papers on Economic Activity: Microeconomics, pp. 1-59, reprinted in: Bailey, E. E. & Rothenberg Pack, J. (eds.) (1995), The Political Economy of Privatization and Deregulation, Cheltenham, UK and Lime, US: Edward Elgar.

Preissl, B. (1998), 'The Regulation of Telecommunication in Europe', in Vierteljahrshefte zur Wirtschaftsforschung, Berlin: Deutsches Institut für Wirtschaftsforschung.

Samuels, W. J. (1992), 'Some Fundamentals of the Economic Role of Government', in Samuels, W. J. (1992), Essays on the Economic Role of Government, vol. 1, Houndmills, Basingstoke: Macmillan.

Skårfors, R. (1997), 'Telegrafverkets inköp av enskilda telefonnät: omstruktureringen av det svenska telefonsystemet 1883-1918', Working Papers in Transport and Communications History, Uppsala: Departments of Economic History, Umeå and Uppsala Universities.

SOU 1992:70, Telelag. Betänkande av Telelagsutredningen, (Report of the Government's Commission on a new Telecommunications Act), Stockholm.

Steinmo, S., Thelen, K., and Longstreth, F. (1992), Structuring Politics, Cambridge: Cambridge University Press,.

Stigler G. J., (1971), 'The Theory of Economic Regulation', Bell Journal of Economics, 2, pp. 3-21.

Telecommunications Act 1993:597.

Thue, L. (1995), 'Electricity Rules - The Formation and Development of Nordic Electricity Regimes', in A. Kaijser & M. Hedin (eds.), Nordic Energy Systems -

Historical Perspectives and Current Issues, Canton, Mass: Science History Publications.

Usher, J. A. (1994), Utilities, Deregulation and Competition Policy, RUSEL Working Paper, No. 19, Exeter: Department of Politics, University of Exeter.

Winston, C. (1993), 'Economic Deregulation: Days of Reckoning for Microeconomists', Journal of Economic Literature, XXXI (3), September, pp. 1263-89, reprinted in: Bailey, E. E. & Rothenberg Pack, J. (eds.) (1995), The Political Economy of Privatization and Deregulation, Cheltenham, UK and Lime, US: Edward Elgar.

Vogel, S. K. (1996), Freer Markets, More Rules, New York, Cornell University Press.